Aircraft Accident Report

**Loss of Control and Impact with Pacific Ocean
Alaska Airlines Flight 261
McDonnell Douglas MD-83, N963AS
About 2.7 Miles North of
Anacapa Island, California
January 31, 2000**

**NTSB/AAR-02/01
PB2002-910402
Notation 7263E
Adopted December 30, 2002**

National Transportation Safety Board
490 L'Enfant Plaza, S.W.
Washington, D.C. 20594

National Transportation Safety Board. 2003. *Loss of Control and Impact with Pacific Ocean, Alaska Airlines Flight 261, McDonnell Douglas MD-83, N963AS, About 2.7 Miles North of Anacapa Island, California, January 31, 2000.* **Aircraft Accident Report NTSB/AAR-02/01. Washington, DC.**

Abstract: This report explains the accident involving Alaska Airlines flight 261, a McDonnell Douglas MD-83, which crashed into the Pacific Ocean about 2.7 miles north of Anacapa Island, California. Safety issues discussed in this report include lubrication and inspection of the jackscrew assembly, extension of lubrication and end play check intervals, jackscrew assembly overhaul procedures, the design and certification of the MD-80 horizontal stabilizer trim control system, Alaska Airlines' maintenance program, and Federal Aviation Administration (FAA) oversight of Alaska Airlines. Safety recommendations are addressed to the FAA.

The National Transportation Safety Board is an independent Federal agency dedicated to promoting aviation, railroad, highway, marine, pipeline, and hazardous materials safety. Established in 1967, the agency is mandated by Congress through the Independent Safety Board Act of 1974 to investigate transportation accidents, determine the probable causes of the accidents, issue safety recommendations, study transportation safety issues, and evaluate the safety effectiveness of government agencies involved in transportation. The Safety Board makes public its actions and decisions through accident reports, safety studies, special investigation reports, safety recommendations, and statistical reviews.

Recent publications are available in their entirety on the Web at <http://www.ntsb.gov>. Other information about available publications also may be obtained from the Web site or by contacting:

**National Transportation Safety Board
Public Inquiries Section, RE-51
490 L'Enfant Plaza, S.W.
Washington, D.C. 20594
(800) 877-6799 or (202) 314-6551**

Safety Board publications may be purchased, by individual copy or by subscription, from the National Technical Information Service. To purchase this publication, order report number **PB2002-910402** from:

**National Technical Information Service
5285 Port Royal Road
Springfield, Virginia 22161
(800) 553-6847 or (703) 605-6000**

The Independent Safety Board Act, as codified at 49 U.S.C. Section 1154(b), precludes the admission into evidence or use of Board reports related to an incident or accident in a civil action for damages resulting from a matter mentioned in the report.

Contents

Abbreviations ix

Executive Summary xii

1. Factual Information 1
  1.1 History of Flight 1
  1.2 Injuries to Persons 10
  1.3 Damage to Airplane 10
  1.4 Other Damage 10
  1.5 Personnel Information 10
    1.5.1 The Captain 10
    1.5.2 The First Officer 11
  1.6 Airplane Information 11
    1.6.1 MD-80 Longitudinal Trim Control System Information 13
      1.6.1.1 Primary Trim Control System 18
      1.6.1.2 Alternate Trim Control System 18
      1.6.1.3 Autopilot Pitch Control 19
    1.6.2 Design and Certification of MD-80 Series Airplanes 19
      1.6.2.1 General 19
      1.6.2.2 Longitudinal Trim Control System Certification 19
    1.6.3 Maintenance Information 24
      1.6.3.1 Alaska Airlines' MD-80 Maintenance Program 24
        1.6.3.1.1 Maintenance Program Development Guidance 24
        1.6.3.1.2 Alaska Airlines' Continuous Airworthiness Maintenance Program 26
        1.6.3.1.3 Alaska Airlines' Reliability Analysis Program 26
      1.6.3.2 Jackscrew Assembly Lubrication Procedures 29
        1.6.3.2.1 Jackscrew Assembly Lubrication Intervals 31
          1.6.3.2.1.1 Manufacturer-Recommended Lubrication Intervals 31
          1.6.3.2.1.2 Alaska Airlines' Lubrication Intervals 32
        1.6.3.2.2 Chronology of Grease Type Changes 34
        1.6.3.2.3 July 19, 2002, Alaska Airlines Maintenance Information Letter 36
      1.6.3.3 Procedures for Monitoring Acme Nut Thread Wear 37
        1.6.3.3.1 Defining and Calculating Wear Rate 37
        1.6.3.3.2 Development of End Play Check Procedures 38
          1.6.3.3.2.1 Alaska Airlines' End Play Check Procedure 43
        1.6.3.3.3 Horizontal Stabilizer Restraining Fixture 44
          1.6.3.3.3.1 General 44
          1.6.3.3.3.2 Alaska Airlines-Manufactured Restraining Fixtures ... 44
        1.6.3.3.4 End Play Check Intervals 47
          1.6.3.3.4.1 Manufacturer-Recommended End Play Check Intervals 47
          1.6.3.3.4.2 Alaska Airlines' End Play Check Intervals 47
      1.6.3.4 Accident Airplane's Maintenance Information 49
        1.6.3.4.1 Accident Airplane's Last C Check Requiring an End Play Check 50

1.6.3.4.2 Maintenance Personnel Statements Regarding the Accident
Airplane's Last End Play Check 53
1.7 Meteorological Information 54
1.8 Aids to Navigation 54
1.9 Communications 54
1.10 Airport Information 55
1.11 Flight Recorders 55
  1.11.1 Cockpit Voice Recorder 55
    1.11.1.1 CVR Sound Spectrum Study 55
  1.11.2 Flight Data Recorder 56
1.12 Wreckage and Impact Information 57
  1.12.1 General 57
  1.12.2 Horizontal Stabilizer Jackscrew Assembly Components 58
1.13 Medical and Pathological Information 62
1.14 Fire 62
1.15 Survival Aspects 62
1.16 Tests and Research 63
  1.16.1 Airplane Performance 63
    1.16.1.1 Airplane Performance Simulation Studies 66
  1.16.2 Metallurgical Examinations 67
    1.16.2.1 Horizontal Stabilizer Forward Spar 67
    1.16.2.2 Acme Screw 68
    1.16.2.3 Acme Screw Torque Tube 70
    1.16.2.4 Lower Mechanical Stop and Corresponding Spline Area on the Acme Screw .. 71
    1.16.2.5 Acme Nut 72
    1.16.2.6 Acme Nut Thread Remnant Examination 76
  1.16.3 Shear Load Capability of the Acme Nut Threads 77
  1.16.4 Studies of Thread Stress and Deformation 77
  1.16.5 Torque Tube Testing 79
  1.16.6 Additional Safety Board Jackscrew Assembly Examinations 80
    1.16.6.1 February 2000 Post-AD 2000-03-51 Jackscrew Assembly Examinations 80
    1.16.6.2 Examinations of Hawaiian Airlines' Jackscrew Assemblies 81
    1.16.6.3 Examinations of Alaska Airlines Reports of Acme Screw "Wobble" 83
  1.16.7 Chemical and Microscopic Analyses of Grease Residues from the Accident
Jackscrew Assembly 84
  1.16.8 Additional Grease Testing, Experiments, and Analysis 85
    1.16.8.1 Standardized Grease Testing 85
    1.16.8.2 Experiments and Analysis on the Surface Chemistry of Acme
Nut Material When Exposed to Various Greases or Grease Mixtures 85
    1.16.8.3 Wear Testing Under Various Grease Conditions 86
1.17 Organizational and Management Information 87
  1.17.1 Alaska Airlines, Inc. 87
  1.17.2 Alaska Airlines Flight Crew Training 88
    1.17.2.1 Stabilizer Trim Check Procedures 88
    1.17.2.2 Runaway Stabilizer Checklist Procedures 89
    1.17.2.3 Stabilizer Inoperative Checklist Procedures 91
      1.17.2.3.1 Postaccident Boeing Flight Operations Bulletin on Stabilizer
Trim Inoperative/Malfunction Procedures 92
  1.17.3 FAA Oversight of Alaska Airlines 93
    1.17.3.1 General 93

1.17.3.2 Preaccident FAA National Aviation Safety Inspection Program Inspection.... 93
1.17.3.3 FAA Air Transportation Oversight System............................93
   1.17.3.3.1 Seattle CMO Memorandum on Staff Shortages for Surveillance
         of Alaska Airlines...94
1.17.3.4 FAA Postaccident Special Inspection of Alaska Airlines..................95
   1.17.3.4.1 FAA Proposed Suspension of Alaska Airlines' Heavy Maintenance
         Authority...98
   1.17.3.4.2 Alaska Airlines Airworthiness and Operations Action Plan............99
   1.17.3.4.3 FAA Followup Evaluation 100
   1.17.3.4.4 DOT Office of the Inspector General Report on FAA Oversight
         of Continuing Analysis and Surveillance Programs..................102
1.17.4 Other Safety Evaluations of Alaska Airlines.............................103
  1.17.4.1 DoD Capability Survey of Alaska Airlines.............................103
  1.17.4.2 Criminal Investigation of Alaska Airlines.............................104
  1.17.4.3 Postaccident Independent Safety Assessment of Alaska Airlines............104
1.18 Additional Information 105
  1.18.1 Postaccident Airworthiness Directives 105
    1.18.1.1 Safety Board and FAA Correspondence Regarding the 2,000-Flight-Hour
        End Play Check Interval Specified in ADs 2000-03-51 and 2000-15-15 106
  1.18.2 Safety Board Statistical End Play Data Study 108
  1.18.3 Previous Safety Recommendations Resulting from the Alaska Airlines
     Flight 261 Investigation...110
  1.18.4 Alaska Airlines Fleetwide MD-80 Jackscrew Assembly Data and Tracking
     History 113
  1.18.5 MD-11 Jackscrew Assembly Acme Nut Wear History.....................116
  1.18.6 Safety Board Observations of Jackscrew Assembly Lubrications.............116
  1.18.7 End Play Check Anomalies...117
    1.18.7.1 Safety Board End Play Check Observations 117
    1.18.7.2 Postaccident Alaska Airlines-Reported Near-Zero End Play Measurements ..118
  1.18.8 Jackscrew Assembly Overhaul Information 119
    1.18.8.1 Jackscrew Overhaul Specifications and Authority 119
    1.18.8.2 Review of the Boeing DC-9 Overhaul Maintenance Manual 120
    1.18.8.3 Safety Board Maintenance Facility Observations 121
    1.18.8.4 Overhauled Jackscrew Assembly Data From 1996 to 1999 122
  1.18.9 Industrywide Jackscrew Assembly Maintenance Procedures 122
  1.18.10 FAA Commercial Airplane Certification Process Study....................123
  1.18.11 Fail-Safe Jackscrew Assembly Designs 125

2. Analysis 126
2.1 General 126
2.2 Accident Sequence 127
  2.2.1 Takeoff and Climbout 127
  2.2.2 Jamming of the Horizontal Stabilizer 127
    2.2.2.1 Cause of the Jam 129
  2.2.3 Release of the Jam and the Initial Dive 132
  2.2.4 The Second and Final Dive 133
  2.2.5 Flight Crew Decision-Making..135
    2.2.5.1 Decision to Continue Flying Rather than Return to PVR 135
    2.2.5.2 Use of the Autopilot...137
    2.2.5.3 Configuration Changes 138

    2.2.5.4   Activation of the Primary Trim Motor. 138
    2.2.5.5   Adequacy of Current Guidance 139
2.3   Evaluation of Potential Reasons for Excessive Acme Nut Thread Wear. 140
   2.3.1   Use of Aeroshell 33 for Lubrication of the Jackscrew Assembly 141
   2.3.2   Acme Screw Thread Surface Finish. 142
   2.3.3   Foreign Debris 142
   2.3.4   Abnormal Loading of the Acme Nut Threads 143
   2.3.5   Summary of Possibilities Ruled Out as Reasons for the Excessive Acme
            Nut Thread Wear 143
   2.3.6   Insufficient Lubrication of the Jackscrew Assembly. 144
      2.3.6.1   Analysis of How Many Recently Scheduled Lubrications Might Have
                 Been Missed or Inadequately Performed Before the Accident 146
      2.3.6.2   Alaska Airlines' Lubrication Interval Extension. 146
        2.3.6.2.1   Safety Implications of Lubrication Interval Extension 147
      2.3.6.3   Adequacy of Lubrication Procedures. 148
2.4   Monitoring Acme Nut Thread Wear. 151
   2.4.1   Alaska Airlines' Preaccident End Play Check Intervals 151
      2.4.1.1   Adequacy of Existing Process for Establishing Maintenance Task Intervals . . 153
   2.4.2   Adequacy of Current End Play Check Intervals 154
   2.4.3   Adequacy of End Play Check Procedure 156
2.5   Deficiencies of Jackscrew Assembly Overhaul Procedures and Practices. 159
2.6   Horizontal Stabilizer Trim System Design and Certification Issues 162
   2.6.1   Acme Nut Thread Loss as a Catastrophic Single-Point Failure Mode 162
   2.6.2   Prevention of Acme Nut Thread Loss Through Maintenance and Inspection. . . . 165
   2.6.3   Elimination of Catastrophic Effects of Acme Nut Thread Loss Through Design . 166
   2.6.4   Consideration of Wear-Related Failures During Design and Certification 167
2.7   Deficiencies in Alaska Airlines' Maintenance Program. 167
   2.7.1   April 2000 FAA Special Inspection Findings 167
   2.7.2   Maintenance-Related Deficiencies Identified During This Accident
            Investigation. 169
      2.7.2.1   General Policy Decisions. 169
      2.7.2.2   Specific Maintenance Actions 172
   2.7.3   Summary 174
2.8   FAA Oversight 174

3. Conclusions 176
3.1   Findings 176
3.2   Probable Cause 180

4. Recommendations. 181
4.1   New Recommendations 181
4.2   Previously Issued Recommendations Resulting From This Accident Investigation . . 184

Board Member Statements 187

5. Appendixes 190

Figures

1. The accident airplane's flightpath, starting about 1609 (about the time of the initial dive) and ending about 1620 (about the time of the second and final dive). 5

2. Radar altitude data and selected ATC transmissions from about 1609 to 1620. 6

3. Installation of jackscrew assembly within the horizontal and vertical stabilizers. 13

4. The MD-80 horizontal and vertical stabilizer tail structure. 14

5. Acme screw and nut. 15

6. Detailed schematic of the longitudinal trim actuating mechanism. 16

7. Cockpit switches, handles, and indicators for the longitudinal control system. 17

8. Depiction of how restraining fixture should be placed on the horizontal stabilizer during an end play check. 40

9. View of a typical dial indicator set up for end play checks. 41

10. Two Alaska Airlines-fabricated restraining fixtures and three Boeing- manufactured fixtures. 46

11. The September 27, 1997, MIG-4. 51

12a. The acme screw immediately after it was brought on board the recovery ship 59

12b. The acme screw immediately after it was brought to shore 60

12c. The acme screw during initial inspection. 61

13. A photograph of the acme screw thread remnants wrapped around the screw, the red rust areas, and the white deposits. 68

14. A photograph of the sand/grease mixture packed between the acme screw's lower threads. 69

15. A photograph of an overall view of the recovered acme nut assembly. 73

16. A photograph of a closer view of the interior of the acme nut. 73

17. A diagram of a cross-section through the acme nut's grease passageway and fitting. Red areas denote the locations where red grease was found, and gray areas denote the location where the dry residue was found. 74

18. A photograph of the grease passageway and counterbore after the acme nut was sectioned. 75

19. A summary of the results of the high-load wear tests, including the contamination and low-temperature tests. 87

20. A graphical depiction of the stages of acme nut thread wear to the point of fracture. 131

Abbreviations

AC	advisory circular
AD	airworthiness directive
ALPA	Air Line Pilots Association
AMFA	Aircraft Mechanics Fraternal Association
AOL	all operators letter
ARTCC	Air Route Traffic Control Center
ASB	alert service bulletin
ATC	air traffic control
ATOS	Air Transportation Oversight System
ATP	airline transport pilot
CAM	cockpit area microphone
CAR	*Civil Aeronautics Regulations*
CFR	*Code of Federal Regulations*
CG	center of gravity
CMO	Certificate Management Office
CMR	certification maintenance requirement
CSP	comprehensive surveillance plan
CVR	cockpit voice recorder
DFGC	digital flight guidance computer
DoD	Department of Defense
DOT	Department of Transportation
EDS	(x-ray) energy dispersive spectroscopy
E.O.	engineering order
FAA	Federal Aviation Administration
FARs	*Federal Aviation Regulations*
FDR	flight data recorder

FEA	finite element analysis
FL	flight level
FSAW	Flight Standards Information Bulletin for Airworthiness
FSDO	Flight Standards District Office
Ftu	minimum ultimate strength
GMM	General Maintenance Manual
KIAS	knots indicated airspeed
LAX	Los Angeles International Airport, Los Angeles, California
MAC	mean aerodynamic chord
MRB	Maintenance Review Board
MSG	Maintenance Steering Group
MSI	maintenance significant item
MTBR	mean time between removal
MTBUR	mean time between unscheduled removal
NASIP	National Aviation Safety Inspection Program
OAK	Oakland International Airport, Oakland, California
OAMP	on-aircraft maintenance planning
OJT	on-the-job training
P/N	part number
PMI	principal maintenance inspector
psi	pounds per square inch
PST	Pacific standard time
PTRS	Program Tracking and Reporting System
PVR	Lic Gustavo Diaz Ordaz International Airport, Puerto Vallarta, Mexico
QRH	Quick Reference Handbook
RMS	root mean squared
S/N	serial number
SAE	Society of Automotive Engineers

SB	service bulletin
SEA	Seattle-Tacoma International Airport, Seattle, Washington
SEM	scanning electron microscopy
SFO	San Francisco International Airport, San Francisco, California
UNS	unified numbering system

Executive Summary

On January 31, 2000, about 1621 Pacific standard time, Alaska Airlines, Inc., flight 261, a McDonnell Douglas MD-83, N963AS, crashed into the Pacific Ocean about 2.7 miles north of Anacapa Island, California. The 2 pilots, 3 cabin crewmembers, and 83 passengers on board were killed, and the airplane was destroyed by impact forces. Flight 261 was operating as a scheduled international passenger flight under the provisions of 14 *Code of Federal Regulations* Part 121 from Lic Gustavo Diaz Ordaz International Airport, Puerto Vallarta, Mexico, to Seattle-Tacoma International Airport, Seattle, Washington, with an intermediate stop planned at San Francisco International Airport, San Francisco, California. Visual meteorological conditions prevailed for the flight, which operated on an instrument flight rules flight plan.

The National Transportation Safety Board determines that the probable cause of this accident was a loss of airplane pitch control resulting from the in-flight failure of the horizontal stabilizer trim system jackscrew assembly's acme nut threads. The thread failure was caused by excessive wear resulting from Alaska Airlines' insufficient lubrication of the jackscrew assembly.

Contributing to the accident were Alaska Airlines' extended lubrication interval and the Federal Aviation Administration's (FAA) approval of that extension, which increased the likelihood that a missed or inadequate lubrication would result in excessive wear of the acme nut threads, and Alaska Airlines' extended end play check interval and the FAA's approval of that extension, which allowed the excessive wear of the acme nut threads to progress to failure without the opportunity for detection. Also contributing to the accident was the absence on the McDonnell Douglas MD-80 of a fail-safe mechanism to prevent the catastrophic effects of total acme nut thread loss.

The safety issues discussed in this report include lubrication and inspection of the jackscrew assembly, extension of lubrication and end play check intervals, jackscrew assembly overhaul procedures, the design and certification of the MD-80 horizontal stabilizer trim control system, Alaska Airlines' maintenance program, and FAA oversight of Alaska Airlines. Safety recommendations are addressed to the FAA.

1. Factual Information

1.1 History of Flight

On January 31, 2000, about 1621 Pacific standard time (PST),[1] Alaska Airlines, Inc., flight 261, a McDonnell Douglas MD-83 (MD-83), N963AS, crashed into the Pacific Ocean about 2.7 miles north of Anacapa Island, California. The 2 pilots, 3 cabin crewmembers, and 83 passengers[2] on board were killed, and the airplane was destroyed by impact forces. Flight 261 was operating as a scheduled international passenger flight under the provisions of 14 *Code of Federal Regulations* (CFR) Part 121 from Lic Gustavo Diaz Ordaz International Airport (PVR), Puerto Vallarta, Mexico, to Seattle-Tacoma International Airport (SEA), Seattle, Washington, with an intermediate stop planned at San Francisco International Airport (SFO), San Francisco, California. Visual meteorological conditions prevailed for the flight, which operated on an instrument flight rules flight plan.

The accident airplane arrived at PVR about 1239 on the day of the accident.[3] The inbound pilots stated that they met the accident pilots outside the airplane and briefly discussed its status.[4]

FDR information from the accident flight indicated that during taxi for takeoff, when the FDR started recording, the horizontal stabilizer[5] was at the 7° airplane-nose-up position, which was the takeoff pitch trim[6] setting. About 1337, the accident airplane departed PVR as flight 261. A National Transportation Safety Board review of FAA air traffic control (ATC) tapes and FDR data from the accident flight indicated that the first

---

[1] Unless otherwise indicated, all times are PST, based on a 24-hour clock. The times reported in the cockpit voice recorder (CVR) transcript were based on a coordinated universal time reference clock used by the Federal Aviation Administration's (FAA) Southern California Terminal Radar Approach Control facility for recorded radar data. CVR times were also correlated to flight data recorder (FDR) data. After the correlation was accomplished, all times from both recorders were converted to PST.

[2] The 83 passengers included 3 children who were less than 2 years old. These children were lap-held passengers.

[3] FDR data indicated that the accident airplane landed at PVR with a horizontal stabilizer angle of 6° airplane nose up.

[4] The inbound captain described the accident flight crew as rested, relaxed, and in good spirits. The inbound first officer stated that the accident captain was happy to see them, upbeat, rested, and ready to go to work.

[5] The MD-80's horizontal stabilizer is a critical flight control located at the top of the vertical stabilizer. The horizontal stabilizer is hinged near its trailing edge so that the leading edge can traverse up and down to provide trim for the airplane in the pitch axis. For a description of the airplane's horizontal stabilizer and other components of the longitudinal trim control system, see section 1.6.1.

[6] Pitch trim (also known as longitudinal trim) is the adjustment of the airplane's horizontal stabilizer to achieve a balanced, or stable, flight condition. The longitudinal trim control system is designed to minimize or eliminate the pilot control forces needed to hold the flight controls in the proper position to achieve the desired pitch for a particular phase of flight.

officer was the pilot flying immediately after the airplane's departure.[7] The flight plan indicated that flight 261's cruising altitude would be flight level (FL) 310.[8]

FDR data indicated that during the initial portion of the climb, the horizontal stabilizer moved at the primary trim motor rate of 1/3° per second from 7° to 2° airplane nose up. According to FDR data, the autopilot was engaged at 1340:12, as the airplane climbed through an altitude of approximately 6,200 feet. Thereafter, the FDR recorded horizontal stabilizer movement at the alternate trim motor rate of 1/10° per second from 2° airplane nose up to 0.4° airplane nose down.[9] At 1349:51, as the airplane continued to climb through approximately 23,400 feet at 331 knots indicated airspeed (KIAS), the CVR recorded the horizontal stabilizer move from 0.25° to 0.4° airplane nose down.[10] This was the last horizontal stabilizer movement recorded until the airplane's initial dive about 2 hours and 20 minutes later. At 1353:12, when the airplane was climbing through 28,557 feet at 296 KIAS, the autopilot disengaged.

FDR information and airplane performance calculations indicated that, during the next 7 minutes, the airplane continued to climb at a much slower rate. During this part of the ascent, the elevators were deflected between -1° and -3°, and the airplane was flown manually using up to as much as 50 pounds of control column pulling force.[11] After reaching level flight, the airplane was flown for about 24 minutes using approximately 30 pounds of pulling force at approximately 31,050 feet and 280 KIAS. The airspeed was then increased to 301 KIAS, and the airplane was flown for almost another 1 hour 22 minutes using about 10 pounds of pulling force. At 1546:59, the autopilot was re-engaged.

According to Alaska Airlines documents, ATC and CVR information, and postaccident interviews with Alaska Airlines dispatch and maintenance personnel, the flight crew contacted the airline's dispatch and maintenance control facilities in SEA some time before the beginning of the CVR transcript at 1549:49[12] to discuss a jammed horizontal stabilizer and a possible diversion to Los Angeles International Airport (LAX), Los Angeles, California.[13] These discussions were conducted on a shared company radio

---

[7] According to FDR data, the No. 2 autopilot was selected during the initial part of the flight. Selection of the No. 2 autopilot commands the No. 2 digital flight guidance computer (DFGC) to provide steering guidance to the first officer's instruments.

[8] FL 310 is 31,000 feet mean sea level, based on an altimeter setting of 29.92 inches of mercury.

[9] Operation of the alternate trim motor during this period is consistent with pitch control being commanded by the autopilot, which was engaged at 1340:12.

[10] KIAS is the speed of the airplane as shown on the airspeed indicator on the cockpit control panel. All airspeed information from the FDR is given in KIAS.

[11] All estimates of pulling force in this report refer to the combined control column forces from both the captain's and the first officer's control columns. It is not known whether these forces were applied by the captain, the first officer, or both.

[12] For a complete transcript of the 31-minute CVR recording, see appendix B.

[13] FDR data indicated that a continuing series of radio transmissions began over the very-high frequency 2 channel (used for all non-ATC radio transmissions) about 1521 and continued until the end of the FDR recording about 1620:56.

frequency between Alaska Airlines' dispatch and maintenance facilities at SEA and its operations and maintenance facilities at LAX.

At 1549:56, the autopilot was disengaged; it was re-engaged at 1550:15. According to the CVR transcript, at 1550:44, SEA maintenance asked the flight crew, "understand you're requesting…diversion to LA…is there a specific reason you prefer LA over San Francisco?"[14] The captain replied, "well a lotta times its windy and rainy and wet in San Francisco and uh, it seemed to me that a dry runway…where the wind is usually right down the runway seemed a little more reasonable."

At 1552:02, an SEA dispatcher provided the flight crew with the current SFO weather (wind was 180° at 6 knots; visibility was 9 miles). The SEA dispatcher added, "if uh you want to land at LA of course for safety reasons we will do that…we'll…tell you though that if we land in LA…we'll be looking at probably an hour to an hour and a half we have a major flow program going right now."[15] At 1552:41, the captain replied, "I really didn't want to hear about the flow being the reason you're calling us cause I'm concerned about overflying suitable airports." At 1553:28, the captain discussed with the first officer potential landing runways at SFO, stating, "one eight zero at six…so that's runway one six what we need is runway one nine, and they're not landing runway one nine." The first officer replied, "I don't think so." At 1553:46, the captain asked SEA dispatch if they could "get some support" or "any ideas" from an instructor to troubleshoot the problem; he received no response. At 1555:00, the captain commented, "it just blows me away they think we're gonna land, they're gonna fix it, now they're worried about the flow, I'm sorry this airplane's [not] gonna go anywhere for a while…so you know." A flight attendant replied, "so they're trying to put the pressure on you," the captain stated, "well, no, yea."

At 1556:08, the SEA dispatcher informed the flight crew that, according to the SFO automatic terminal information service, the landing runways in use at SFO were 28R and 28L and that "it hasn't rained there in hours so I'm looking at…probably a dry runway." At 1556:26, the captain stated that he was waiting for a requested center of gravity (CG) update (for landing), and then he requested information on wind conditions at LAX. At 1556:50, the SEA dispatcher replied that the wind at LAX was 260° at 9 knots.

Nine seconds later, the captain, comparing SFO and LAX wind conditions, told the SEA dispatcher, "versus a direct crosswind which is effectively no change in groundspeed…I gotta tell you, when I look at it from a safety point I think that something that lowers my groundspeed makes sense."[16] The SEA dispatcher replied, "that'll mean LAX then for you." He then asked the captain to provide LAX operations with the information needed to recompute the airplane's CG because "they can probably whip out that CG for you real quick." At 1558:15, the captain told the SEA dispatcher, "we're goin

---

[14] This is the first reference on the CVR recording to the accident flight crew's request to divert to LAX.

[15] The SEA dispatcher's comment refers to ongoing departure delays at LAX caused by heavy air traffic congestion.

[16] If the airplane had landed at SFO, it would have encountered a direct crosswind from the south. If the airplane had landed at LAX, it would have encountered a minimal crosswind from the southwest.

to LAX we're gonna stay up here and burn a little more gas[17] get all our ducks in a row, and then we'll uh be talking to LAX when we start down to go in there." At 1558:45, the captain asked LAX operations if it could "compute [the airplane's] current CG based on the information we had at takeoff."

At 1602:33, the captain asked LAX operations for wind information at SFO. LAX operations replied that the winds at SFO were 170° at 6 knots. The captain replied, "that's what I needed. We are comin in to see you." At 1603:56, the first officer began giving LAX operations the information it needed to recompute the airplane's CG for landing.

At 1607:54, a mechanic at Alaska Airlines' LAX maintenance facility contacted the flight crew on the company radio frequency and asked, "are you [the] guys with the uh, horizontal [stabilizer] situation?" The captain replied, "affirmative," and the mechanic, referring to the stabilizer's primary trim system, asked, "did you try the suitcase handles and the pickle switches?"[18] At 1608:03, the captain replied, "yea we tried everything together." At 1608:08, the captain added, "we've run just about everything if you've got any hidden circuit breakers we'd love to know about 'em."[19] The mechanic stated that he would "look at the uh circuit breaker uh guide just as a double check." The LAX mechanic then asked the flight crew about the status of the alternate trim system, and, at 1608:35, the captain replied that "it appears to be jammed...the whole thing, it [the AC load meter] spikes out when we use the primary, we get AC [electrical] load that tells me the motor's tryin to run but the brake won't move it. when we use the alternate, nothing happens."[20]

At 1608:50, the LAX mechanic asked, "you say you get a spike...on the meter up there in the cockpit when you uh try to move it with the...primary right?" According to the CVR transcript, at 1608:59, the captain addressed the first officer before responding to the mechanic, stating, "I'm gonna click it off you got it." One second later, the first officer replied, "ok." At 1609:01, the captain reiterated to the LAX mechanic that the spike occurred "when we do the primary trim but there's no appreciable uh change in the uh electrical uh when we do the alternate." The LAX mechanic replied that he would see them when they arrived at the LAX maintenance facility.

---

[17] The airplane did not have an in-flight fuel dumping system.

[18] "Suitcase handles" is a colloquial term for the longitudinal trim handles located on the center control pedestal. "Pickle switches" is a colloquial term for the trim switches located on the outboard side of each of the control wheels.

[19] According to the CVR transcript, at 1550:14, the first officer had asked the captain to move his seat forward so that he could check a panel behind the captain's seat that contained circuit breakers for the horizontal stabilizer trim system. The first officer then stated to the captain, "I don't think there's anything beyond that we haven't checked."

[20] According to postaccident testing, the indicator needle on the AC load meter in the cockpit normally "spikes," or rises, to a maximum indication when the primary trim motor is engaged. Once the acme screw begins to rotate, the amount of electrical current required to sustain the rotation drops significantly, and the indicator needle will quickly drop to near zero. However, if the acme screw is held fixed or jammed while the primary trim motor is engaged, the indicator needle will remain at the maximum indication. Engagement of the alternate trim motor, regardless of the acme screw's condition, does not cause the indicator needle to spike because of its significantly lower electrical current requirements. For more information about the primary and alternate trim systems, see sections 1.6.1.1 and 1.6.1.2.

At 1609:13, the captain stated, "lets do that." At 1609:14.8, the CVR recorded the sound of a click and, at the same time, the captain stating, "this'll click it off." According to FDR data, the autopilot was disengaged at 1609:16. At the same time, the CVR recorded the sound of a clunk, followed by two faint thumps in short succession at 1609:16.9. The CVR recorded a sound similar to the horizontal stabilizer-in-motion tone[21] at 1609:17. At 1609:19.6, the CVR again recorded a sound similar to the horizontal stabilizer-in-motion tone, followed by the captain's comment, "you got it?" (FDR data indicated that during the 3 to 4 seconds after the autopilot was disengaged, the horizontal stabilizer moved from 0.4° to a recorded position of 2.5° airplane nose down, and the airplane began to pitch nose down, starting a dive that lasted about 80 seconds as the airplane went from about 31,050 to between 23,000 and 24,000 feet.)[22] Figure 1 shows the accident airplane's flightpath, starting about 1609 (about the time of the initial dive) and ending about 1620 (about the time of the second and final dive). Figure 2 shows the radar altitude data and selected ATC transmissions for about the same time period.

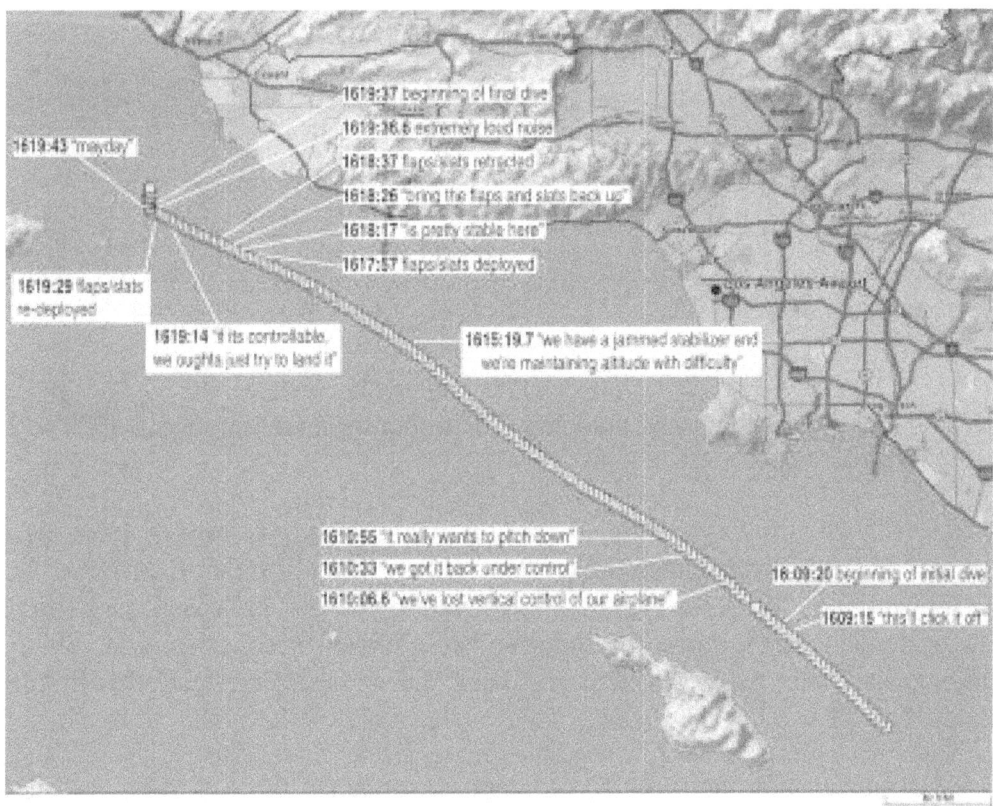

Figure 1. The accident airplane's flightpath, starting about 1609 (about the time of the initial dive) and ending about 1620 (about the time of the second and final dive).

---

[21] The MD-80 is equipped with a sensor to detect movement of a cable loop that is connected to the horizontal stabilizer. If the horizontal stabilizer continuously moves through about 1.1°, a 1-second audible signal is produced in the cockpit by the sensor. This audible signal is produced about every 0.55° of continuous movement thereafter. The sensor resets when the horizontal stabilizer stops moving.

[22] During the initial dive, the airplane accelerated to a maximum speed of 353 KIAS, and this value remained nearly constant until 1610:23 as the airplane was descending through 24,200 feet.

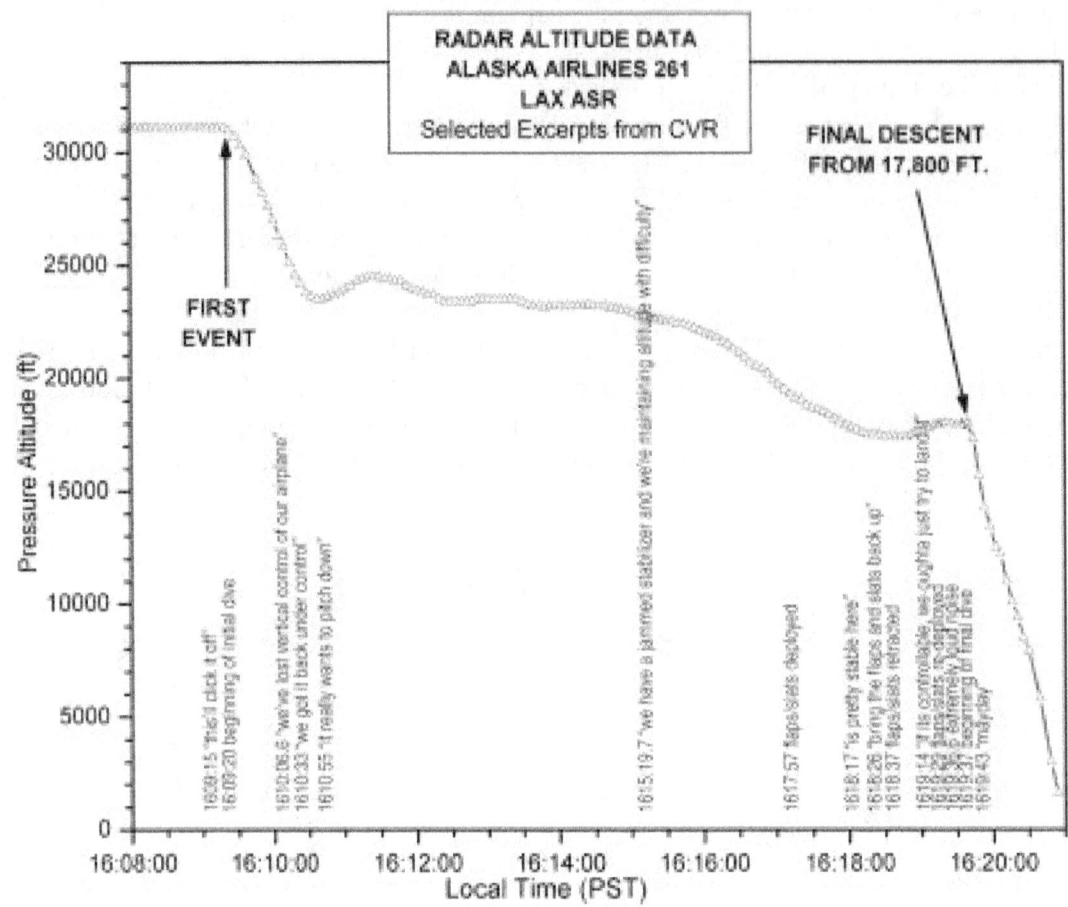

Figure 2. Radar altitude data and selected ATC transmissions from about 1609 to 1620.

At 1609:26, the captain stated, "it got worse," and, 5 seconds later, he stated "you're stalled." One second later, the CVR recorded a sound similar to airframe vibration getting louder. At 1609:33, the captain stated, "no no you gotta release it ya gotta release it." This statement was followed by the sound of a click 1 second later. At 1609:52, the captain stated, "help me back help me back." Two seconds later, the first officer responded, "ok."

One second later, at 1609:55, the captain contacted the Los Angeles Air Route Traffic Control Center (ARTCC)[23] and stated, "center Alaska two sixty one we are uh in a dive here." At 1610:01.6, the captain added, "and I've lost control, vertical pitch." At 1610:01.9, the CVR recorded the sound of the overspeed warning (which continued for the next 33 seconds). At 1610:05, the controller asked flight 261 to repeat the transmission, and, at 1610:06, the captain responded, "yea we're out of twenty six

---

[23] According to the CVR and ATC transcripts, flight 261 had established contact with the Los Angeles ARTCC controller about 1551:21.

thousand feet, we are in a vertical dive...not a dive yet...but uh we've lost vertical control of our airplane." At 1610:20, the captain stated, "just help me."

At 1610:28.2, the captain informed the Los Angeles ARTCC controller, "we're at twenty three seven request uh." At 1610:33, the captain added, "yea we got it back under control here." One second later, the first officer transmitted, "no we don't." At 1610:45, the first officer stated, "let's take the speedbrakes off." One second later, the captain responded, "no no leave them there. it seems to be helping." At 1610:55, the captain stated, "ok it really wants to pitch down." At 1611:06.6, the captain stated that they were at "twenty four thousand feet, kinda stabilized." Three seconds later he added, "we're slowin' here, and uh, we're gonna uh do a little troubleshooting, can you gimme a block [altitude] between uh, twenty and twenty five?"[24] FDR data indicated that, by 1611:13, the airplane's airspeed had decreased to 262 KIAS, and the airplane was maintaining an altitude of approximately 24,400 feet with a pitch angle of 4.4°. At 1611:21, the controller assigned flight 261 a block altitude of between FL 200 and 250. Airplane performance calculations indicated that between about 130 and 140 pounds of pulling force was required to recover from the dive.

At 1611:43, the first officer stated, "whatever we did is no good, don't do that again." One second later, the captain responded, "yea, no it went down it went to full nose down." Four seconds later, the first officer asked, "uh it's a lot worse than it was?" At 1611:50, the captain replied, "yea yea we're in much worse shape now," adding, at 1611:59, "I think it's at the stop, full stop...and I'm thinking...can it go any worse...but it probably can...but when we slowed down, lets slow it lets get down to two hundred knots and see what happens."

At 1612:33, the captain told LAX maintenance, "we did both the pickle switch and the suitcase handles and it ran away full nose trim down." At 1612:42, the captain added, "and now we're in a pinch so we're holding uh we're worse than we were." At 1613:04, the captain indicated to LAX maintenance that he was reluctant to try troubleshooting the trim system again because the trim might "go in the other direction." At 1613:10, the LAX mechanic responded, "ok well your discretion uh if you want to try it, that's ok with me if not that's fine. um we'll see you at the gate." At 1613:22, the captain stated, "I went tab down...right, and it should have come back instead it went the other way." At 1613:32, the captain asked the first officer, "you wanna try it or not?" The first officer replied, "uhh no. boy I don't know." Airplane performance calculations indicated that about 120 pounds of pulling force was being applied to the pilots' control columns at this point.

At 1614:54, the Los Angeles ARTCC controller instructed the flight crew to contact another ARTCC controller on frequency 126.52, which the flight crew acknowledged. At 1615:19, the first officer contacted another ARTCC controller on 126.52 and stated, "we're with you we're at twenty two five, we have a jammed stabilizer and we're maintaining altitude with difficulty. uh but uh we can maintain altitude we think...our intention is to land at Los Angeles." The controller cleared the airplane direct

---

[24] A block altitude assignment allows an airplane to operate between the assigned upper and lower altitude limits.

to LAX and then asked, "you want lower [altitude] now or what do you want to do sir?" At 1615:56, the captain replied, "I need to get down about ten, change my configuration, make sure I can control the jet and I'd like to do that out here over the bay if I may."

At 1616:32, the Los Angeles ARTCC controller issued flight 261 a heading of 280° and cleared the flight to descend to 17,000 feet. At 1616:39, the captain acknowledged, "two eight zero and one seven seventeen thousand Alaska two sixty one. and we generally need a block altitude." At 1616:45, the controller responded, "ok and just um I tell you what do that for now sir, and contact LA center on one three five point five they'll have further uh instructions for you sir." At 1616:56.9, the first officer acknowledged, "ok thirty five five say the altimeter setting." The controller responded, "the LA altimeter is three zero one eight." At 1617:02, the first officer responded, "thank you." According to the CVR and ATC recordings, this was the last radio transmission made from flight 261.

After the radio transmission, the captain told a flight attendant that he needed "everything picked up" and "everybody strapped down." At 1617:04, the captain added, "I'm gonna unload the airplane and see if we can...we can regain control of it that way." At 1617:09, the flight attendant stated, "ok we had like a big bang back there," and, the captain replied, "yea I heard it." At 1617:15, the captain stated, "I think the stab trim thing is broke." At 1617:21, the captain again told the flight attendant to make sure the passengers were "strapped in now," adding 3 seconds later, "cause I'm gonna I'm going to release the back pressure and see if I can get it...back."

At 1617:54, the captain stated, "gimme slats extend," and, at 1617:56.6, a sound similar to slat/flap handle movement was recorded by the CVR. At 1617:58, the captain added, "I'm test flyin now." At 1618:05, the captain commanded an 11° flap deployment, and, at 1618:07, a sound similar to slat/flap handle movement was recorded. At 1618:17, the captain stated, "its pretty stable right here...see but we got to get down to a hundred an[d] eighty [knots]." At 1618:26, the captain stated, "OK...bring bring the flaps and slats back up for me," and, at 1618:36.8, sounds similar to slat/flap handle movement were recorded. At 1618:47, the captain stated, "what I wanna do...is get the nose up...and then let the nose fall through and see if we can stab it when it's unloaded."

At 1618:56, the first officer responded, "you mean use this again? I don't think we should...if it can fly." At 1619:01, the captain replied, "it's on the stop now, it's on the stop." At 1619:04, the first officer replied, "well not according to that it's not." At this time, FDR data indicated a horizontal stabilizer angle of 2.5° airplane nose down. Three seconds later, the first officer added, "the trim might be, and then it might be uh, if something's popped back there... it might be mechanical damage too." At 1619:14, the first officer stated, "I think if it's controllable, we oughta just try to land it." Two seconds later, the captain replied, "you think so? ok lets head for LA."

About 5 seconds later, the CVR recorded the sound of a series of at least four distinct "thumps."[25] At 1619:24, the first officer asked, "you feel that?" and the captain

---

[25] The CVR transcript describes this sound as a "faint thump."

replied, "yea." At 1619:29, the captain stated, "ok gimme sl---." At 1619:32.8, the CVR recorded the sound of two clicks similar to the sound of slat/flap movement. At 1619:36.6, the CVR recorded the sound of an "extremely loud noise" and the sound of background noise increasing, which continued until the end of the recording. At the same time, the CVR also recorded sounds similar to loose articles moving around the cockpit. FDR data indicated that at 1619:36.6, the flaps were extending and the slats were moving to the mid position. The next few seconds of FDR data indicated a maximum airplane-nose-down pitch rate of nearly 25° per second. The FDR recorded a significant decrease in vertical acceleration values (negative Gs),[26] a nose-down pitch angle, and a significant decrease in lateral acceleration values. By 1619:40, the airplane was rolling left wing down, and the rudder was deflected 3° to the right.

FDR data indicated that, by 1619:42, the airplane had reached its maximum valid recorded airplane-nose-down pitch angle of -70°. At this time, the roll angle was passing through -76° left wing down. At 1619:43, the first officer stated, "mayday," but did not make a radio transmission. Six seconds later, the captain stated, "push and roll, push and roll." FDR data indicated that, by 1619:45, the pitch angle had increased to -28°, and the airplane had rolled to -180° (inverted). Further, the airplane had descended to 16,420 feet, and the indicated airspeed had decreased to 208 knots.

At 1619:54, the captain stated, "ok, we are inverted...and now we gotta get it." FDR data indicated that at this time, the left aileron moved to more than 16° (to command right wing down), then, during the next 6 seconds, it moved in the opposite direction to -13° (to command left wing down). At 1619:57, the rudder returned to the near 0° position, the flaps were retracted, and the airplane was rolling through -150° with an airplane-nose-down pitch angle of -9°. After 1619:57, the airplane remained near inverted and its pitch oscillated in the nose-down position.

At 1620:04, the captain stated, "push push push...push the blue side up." At 1620:16, the captain stated, "ok now lets kick rudder...left rudder left rudder." Two seconds later, the first officer replied, "I can't reach it." At 1620:20, the captain replied, "ok right rudder...right rudder." At 1620:38, the captain stated, "gotta get it over again...at least upside down we're flyin." At 1620:49, the CVR recorded sounds similar to engine compressor stalls and engine spooldown.[27] At 1620:54, the captain commanded deployment of the speedbrakes, and, about 1 second later, the first officer replied, "got it." At 1620:56.2, the captain stated, "ah here we go." The FDR recording ended at 1620:56.3, and the CVR recording ended at 1620:57.1.

The airplane impacted the Pacific Ocean near Port Hueneme, California. Pieces of the airplane wreckage were found floating on and beneath the surface of the ocean. The main wreckage was found at 34° 03.5' north latitude and 119° 20.8' west longitude.

---

[26] A G is a unit of measurement of force on a body undergoing acceleration as a multiple of its weight. The normal load factor for an airplane in straight and level flight is about 1 G. As the load factor decreases from 1 G, objects become increasingly weightless, and at 0 G, those objects float.

[27] An engine compressor stall, or surge, results from interruption of normal airflow through the engine, which can be caused by an engine malfunction or disturbance of inlet airflow at high angles-of-attack.

1.2 Injuries to Persons

Table 1. Injury chart.

Injuries	Flight Crew	Cabin Crew	Passengers	Other	Total
Fatal	2	3	83	0	88
Serious	0	0	0	0	0
Minor	0	0	0	0	0
None	0	0	0	0	0
Total	2	3	83	0	88

1.3 Damage to Airplane

The airplane was destroyed by impact forces.

1.4 Other Damage

No other damage resulted from this accident.

1.5 Personnel Information

The accident flight crew spent the night before the accident at Alaska Airlines' layover hotel, which was in a resort located outside of Puerto Vallarta. Witness statements and hotel records indicated that the captain and first officer watched the Super Bowl in the hotel lounge and ate both breakfast and dinner in the hotel restaurant.[28] The flight crew that flew the accident airplane into PVR spoke to the accident flight crewmembers, and they indicated that the flight crew appeared rested before the accident flight.

1.5.1 The Captain

The captain, age 53, was hired by JetAmerica[29] on August 16, 1982. The captain held an airline transport pilot (ATP) certificate, issued March 16, 1984, with a type rating

---

[28] For more information about the flight crew's preaccident activities, see the Operations/Human Performance Group Chairman's report and addendums in the Safety Board's public docket for this accident.

[29] JetAmerica merged with Alaska Airlines in 1987. According to Alaska Airlines records, the captain began JetAmerica merger training on September 8, 1987.

in the Douglas DC-9 (DC-9)/McDonnell Douglas MD-80 (MD-80). Additionally, he held a turbojet flight engineer certificate and ground and flight instructor certificates. The captain's most recent FAA first-class medical certificate was issued on November 15, 1999, and contained the limitation that he must wear corrective lenses.

According to Alaska Airlines records, the captain had flown approximately 17,750 total flight hours, including 10,460 hours as pilot-in-command, of which about 4,150 hours were as pilot-in-command in the MD-80. He had flown approximately 133, 52, and 24 hours in the last 90, 30, and 7 days, respectively, before the accident. According to company records, the captain satisfactorily completed his last line check on July 15, 1999; his last recurrent training on November 19, 1999; and his last proficiency check on November 23, 1999. Company records also indicated that the captain was off-duty for 2 days before the day of the accident flight. A search of FAA and company records showed no accident, incident, or enforcement or disciplinary actions, and a search of records at the National Driver Register found no history of driver's license revocation or suspension.

1.5.2 The First Officer

The first officer, age 57, was hired by Alaska Airlines on July 17, 1985.[30] The first officer held an ATP certificate, issued April 3, 1985, with type ratings in the DC-9/MD-80, DC-6, and DC-7. The flight officer's most recent FAA second-class medical certificate was issued on April 7, 1999, and contained the limitation that he must wear corrective lenses.

According to Alaska Airlines records, the first officer had flown approximately 8,140 total flight hours, including about 8,060 hours as first officer in the MD-80. He had flown approximately 142, 78, and 15 hours in the past 90, 30, and 7 days, respectively, before the accident. According to company records, the first officer satisfactorily completed his last line check on March 16, 1997; his last recurrent training on April 23, 1999; and his last proficiency check on April 25, 1999. Company records also indicated that the first officer was off-duty for 3 days before the day of the accident flight. A search of FAA and company records showed no accident, incident, or enforcement or disciplinary actions, and a search of records at the National Driver Register found no history of driver's license revocation or suspension.

1.6 Airplane Information

The MD-80 is a low-wing, twin engine, transport-category airplane. The MD-80, which the FAA certified in August 1980, was derived from earlier DC-9 series model airplanes. As a result, much of the MD-80's structure and many of its systems, components, and installations are similar to the earlier DC-9 model. According to

---

[30] According to Alaska Airlines records, the first officer also began JetAmerica merger training on September 8, 1987.

Boeing,[31] the first DC-9 airplane entered service in December 1965; the final DC-9 airplane entered service in October 1982. MD-80 series model airplanes—the MD-81, -82, -83, and -88—were in production through 1999.[32] The DC-9 family of airplanes also includes McDonnell Douglas MD-90 (MD-90) and Boeing 717 (717) series airplanes.[33]

Alaska Airlines began operating MD-80 airplanes on June 30, 1984. The accident airplane, N963AS, an MD-83, serial number (S/N) 53077, was manufactured in 1992 and added to Alaska Airlines' operating certificate on May 27, 1992.[34] According to Alaska Airlines records, the airplane had accumulated about 26,584 total hours of operation (14,315 flight cycles)[35] at the time of the accident. The airplane was configured to seat a maximum of 12 first-class and 128 economy-class passengers and to carry cargo.

The accident airplane was equipped with two Pratt & Whitney JT8D-217C turbofan engines. Company records indicated that the No. 1 (left) engine, S/N 728068, was manufactured in May 1995 and installed on the accident airplane on April 20, 1999, and had operated about 16,970 hours (8,994 flight cycles) since new. At the time of the accident, the No. 1 engine had 3,006 flight cycles remaining until a limited inspection of its low-pressure turbine shaft was required. The No. 2 (right) engine, S/N 726852, was manufactured in April 1992 and installed on the accident airplane on May 24, 1999, and had operated about 24,034 hours (13,266 flight cycles) since new. At the time of the accident, the No. 2 engine had 6,305 flight cycles remaining until a limited inspection of its low-pressure turbine shaft was required.

According to Alaska Airlines dispatch documents for the accident flight, the accident airplane's takeoff weight was calculated to be 136,513 pounds,[36] including 34,902 pounds of fuel. Alaska Airlines' load sheet for the accident flight indicated that there were 10 passengers in the first-class cabin, 70 passengers (not including 3 children)[37] in the coach cabin, and 5 crewmembers (2 flight crewmembers and 3 cabin crewmembers) on board the airplane. The dispatch documents indicated that the airplane's takeoff CG was calculated to be 12.8 percent mean aerodynamic chord (MAC).[38]

---

[31] The Boeing Commercial Airplane Group and the McDonnell Douglas Corporation merged in August 1997. The Douglas Aircraft Company became the McDonnell Douglas Corporation in April 1967 when it merged with the McDonnell Aircraft Company.

[32] A total of 1,191 MD-80 series airplanes were produced.

[33] A total of 117 MD-90 series airplanes were produced from February 13, 1993, to October 23, 2000. A total of 104 717s have been produced, starting on September 23, 1999, and 717s are still in production.

[34] According to Alaska Airlines, the registered airplane owner was the First Security Bank of Utah.

[35] A flight cycle is one complete takeoff and landing sequence.

[36] The accident airplane's weight and balance were calculated using the January 10, 2000, revision of the Alaska Airlines MD-80 Loading Handbook. The maximum certificated takeoff gross weight for the accident airplane was 160,000 pounds, the maximum climb weight for takeoff was 153,900 pounds, and the maximum landing weight was 130,000 pounds.

[37] According to the Alaska Airlines MD-80 Loading Handbook, passengers under 2 years of age are not considered for weight and balance purposes during normal operations.

[38] The maximum forward CG limit for the accident airplane was 8.0 percent MAC, and the maximum aft CG limit was 26.0 percent MAC.

1.6.1 MD-80 Longitudinal Trim Control System Information

Longitudinal control for the MD-80 (and for all DC-9, MD-90, and 717 series airplanes), that is, control of the airplane's pitch movements, is provided by the horizontal stabilizer and the elevators. The horizontal stabilizer is mounted on top of the 18-foot-high vertical stabilizer; they are connected by two hinges at the aft spar of the horizontal stabilizer and with a single jackscrew assembly at the front spar of the stabilizer in a T-tail configuration (see figure 3). The horizontal stabilizer is about 40 feet long and comprises a center box and a left and a right outboard section. Each outboard section of the horizontal stabilizer has an elevator hinged to its trailing edge. Coarse pitch movements are achieved with elevator movement via mechanical linkage from the elevator control tabs to the cockpit control columns (see figure 4). Finer adjustments to pitch are achieved by changing the angle of the entire horizontal stabilizer. The leading edge of the horizontal stabilizer is raised or lowered by the jackscrew assembly as the stabilizer's trailing edge pivots (rotates) about its hinge points.

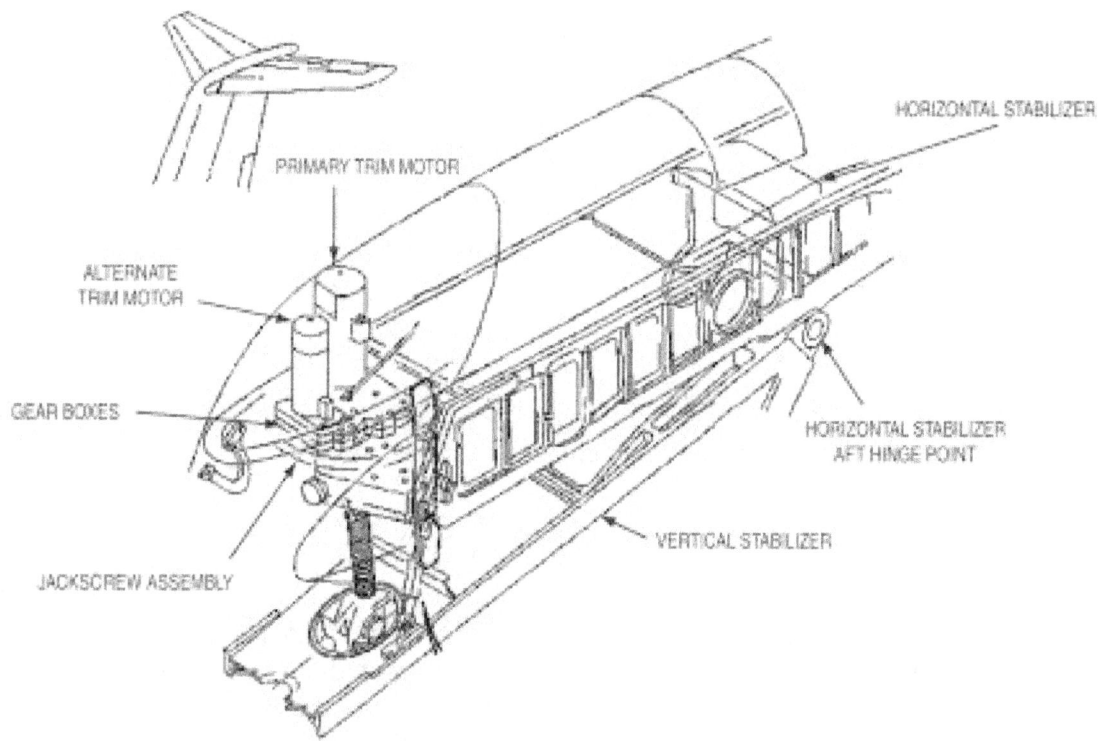

Figure 3. Installation of jackscrew assembly within the horizontal and vertical stabilizers.

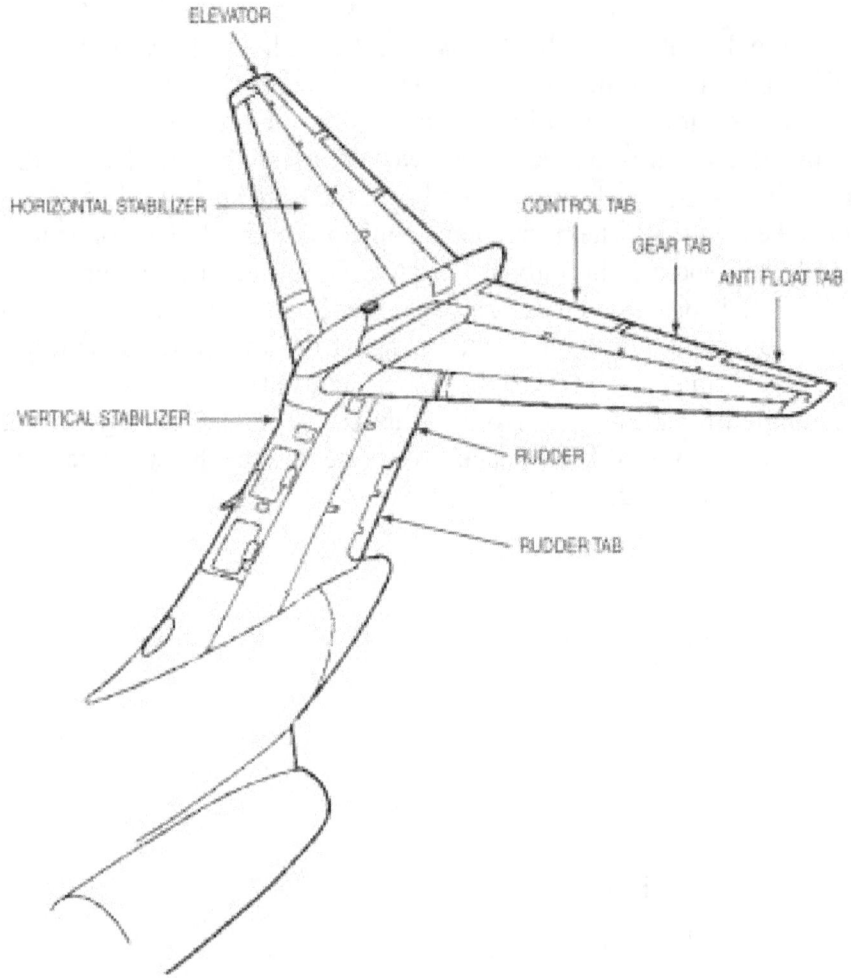

Figure 4. The MD-80 horizontal and vertical stabilizer tail structure.

Movement of the horizontal stabilizer is provided by the jackscrew assembly, which consists of an acme screw and nut, a torque tube inside the acme screw, two gearboxes, two trim motors (an alternate and a primary), and associated components and supports. (See figures 3 through 6 for details of the longitudinal trim system actuating mechanism and the jackscrew assembly.) The upper end of the jackscrew assembly is attached to the front spar of the horizontal stabilizer, and the lower end is threaded through the acme nut, which is attached to the vertical stabilizer with a gimbal ring and retaining pins. The acme screw and nut each have two threads that rotate in a spiral along their length. Figure 6 shows the longitudinal trim system and the jackscrew assembly.

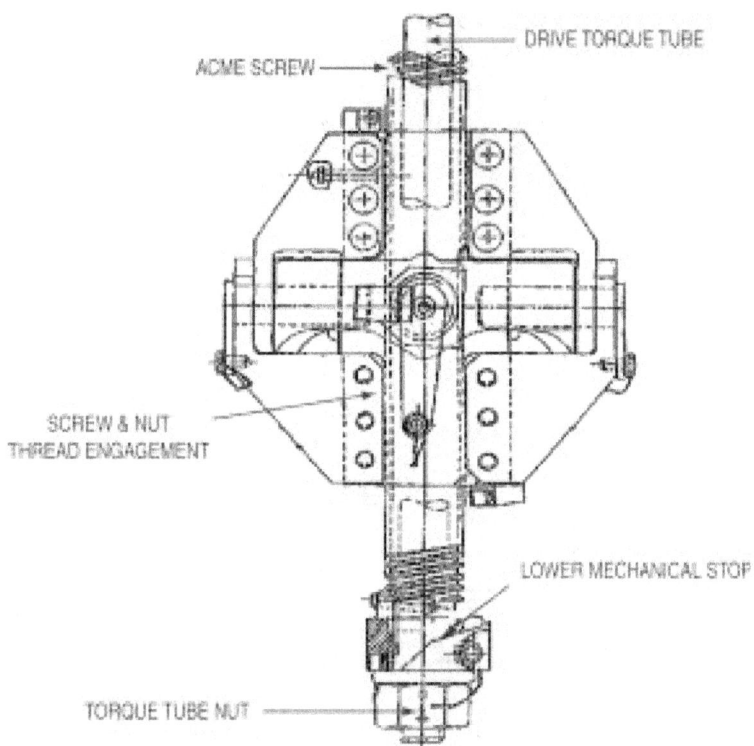

Figure 5. Acme screw and nut.

Movement of the horizontal stabilizer is commanded either automatically by the autopilot when it is engaged, or manually by the flight crew by depressing either set of dual trim switches (located on each control wheel), moving the dual longitudinal trim handles on the center control pedestal, or moving the dual alternate trim control switches on the center pedestal (see figure 7). Any of these commands activates one of the two electric motors that rotate the acme screw by applying torque to the titanium torque tube that is held fixed inside the acme screw. The motors are deenergized whenever either the autopilot senses that the horizontal stabilizer has reached the desired pitch trim condition, when pilot commands are terminated, or when the horizontal stabilizer reaches its maximum travel limits. Electrical travel limit shutoff switches (also known as the electrical stops) stop the motors at the maximum limits of travel. The MD-80 horizontal stabilizer's design limits are 12.2° leading edge down, which results in airplane-nose-up trim, and 2.1° leading edge up, which results in airplane-nose-down trim, as set by the electrical stops.

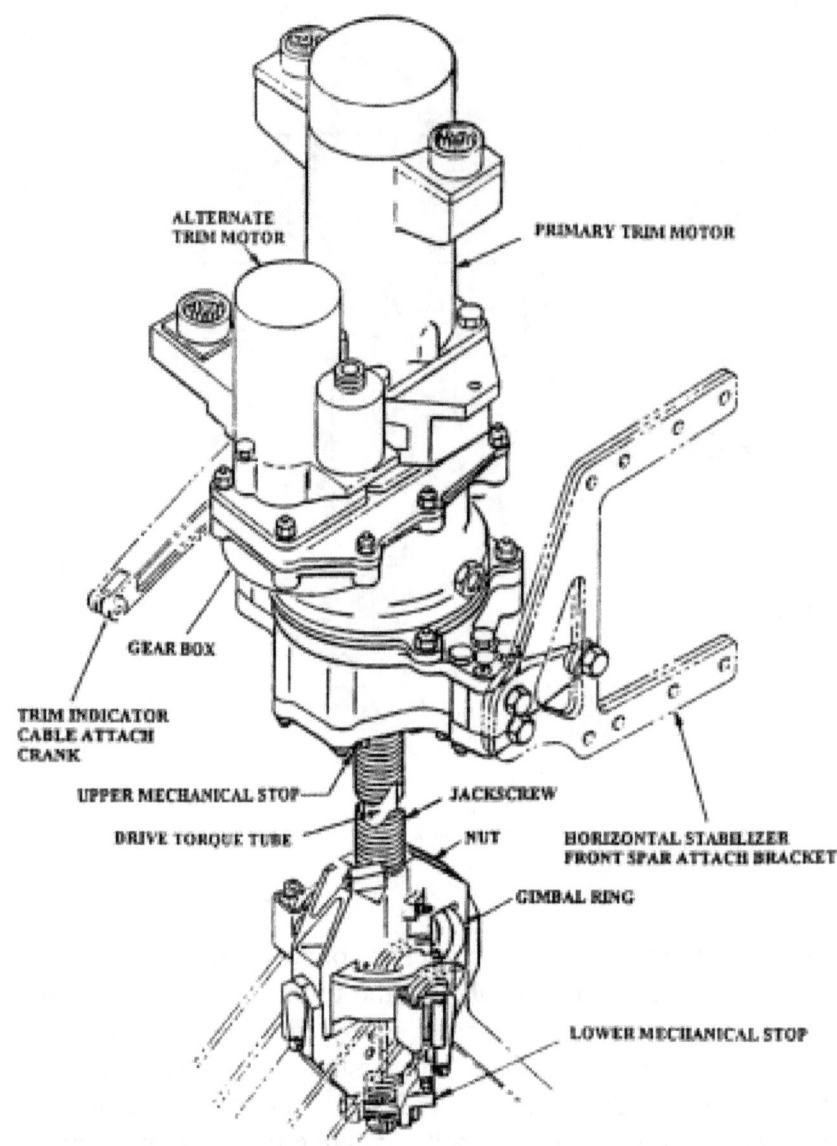

Figure 6. Detailed schematic of the longitudinal trim actuating mechanism.

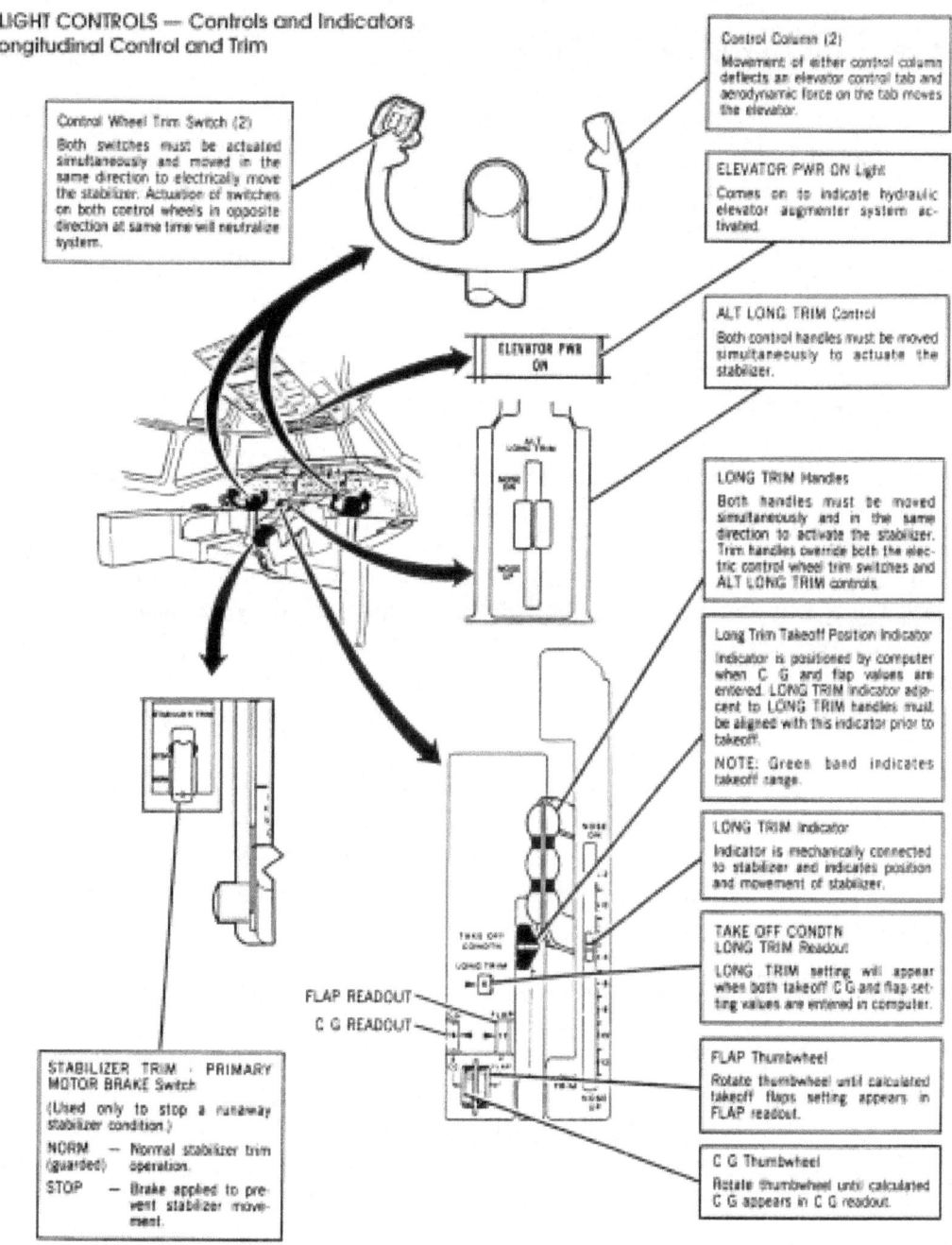

Figure 7. Cockpit switches, handles, and indicators for the longitudinal control system.

The jackscrew assembly also has upper and lower mechanical stops splined to the acme screw to stop screw rotation if the travel limit shutoff switches malfunction. The mechanical stops rotate with the acme screw as it turns and act as the mechanical limits for its range of travel.[39] Stopping action is provided by contact between the head of a steel stop bolt in the mechanical stop and the head of a similar bolt in the acme nut.

1.6.1.1 Primary Trim Control System

As previously stated, the primary trim control system is activated by depressing either set of control wheel trim switches or moving the dual longitudinal trim handles. The pair of switches located on each control wheel is designed to be depressed simultaneously. Both switches on the same control wheel must be moved simultaneously in the same direction to command up or down movement of the horizontal stabilizer. One switch is a motor control switch; the other switch is a brake control switch. Moving the motor control switch energizes the airplane-nose-up or nose-down circuits. Moving the brake control switch energizes the corresponding brake control relay and releases the brake in the trim motor. Energizing the brake control relay automatically disengages the autopilot.

The pair of longitudinal trim handles on the left side of the control pedestal also serve as a mechanical means of activating the trim control system in the event of an electrical circuit failure in the control wheel switches. The pair of handles is designed to be gripped with one hand and moved simultaneously in the same direction. Operation of the handles in a trim direction opposite that being commanded by the control wheel trim switches will override the control wheel trim switch command. A primary trim brake switch is also installed on the center pedestal; it is designed to stop horizontal stabilizer movement, when activated, if a malfunction occurs in the primary trim control system.

The primary trim control system uses the primary trim motor, which is mounted on top of the jackscrew assembly. When powered by the primary trim motor, the horizontal stabilizer moves about 1/3° per second. The primary trim motor rotates the acme screw at speeds of about 35 rpm and has a maximum torque capability (stall torque) of 18,850 inch-pounds.

1.6.1.2 Alternate Trim Control System

As previously mentioned, the alternate trim control system is activated by autopilot commands or by activation of a pair of alternate trim control switches, which are located in the center of the control pedestal, aft of the throttles. These alternate trim control switches must be moved simultaneously in the same direction to activate the alternate trim system. As with the primary trim system, one switch controls the motor, and the other switch controls the brake.

The alternate trim control system uses the alternate trim motor, which is mounted on the jackscrew assembly next to the primary trim motor. When powered by the alternate trim motor, the horizontal stabilizer moves about 1/10° per second. The alternate trim

---

[39] Boeing engineering data, computer modeling, and ground demonstrations on an actual airplane indicated that the maximum possible horizontal stabilizer position when the acme screw lower mechanical stop is pulling up on, but is restrained by, the acme nut is about 3.1° airplane nose down.

motor rotates the acme screw at speeds of about 10 rpm and has a maximum torque capability of 4,400 inch-pounds.

1.6.1.3 Autopilot Pitch Control

When the autopilot is engaged and the flight crew selects a desired flight mode (for example, maintaining a constant altitude, pitch angle, or rate of climb), the autopilot will transmit pitch control signals to the DFGC. The DFGC processes the signals and sends them to the elevator control tab servos, which move the elevator control tabs the proper amount to execute the commanded pitch maneuver. If the signals to the servos are sustained for more than 5 seconds, the DFGC will then command the alternate trim motor to move the horizontal stabilizer in a direction that will relieve the elevator deflection.

Autopilot trim annunciator warning lights in the cockpit illuminate to alert the flight crew if a horizontal stabilizer out-of-trim condition exists while the autopilot is engaged. Specifically, after 10 seconds if there is no horizontal stabilizer movement in response to autopilot commands, the autopilot trim annunciator warning lights illuminate. If such a condition exists, the autopilot will not automatically disengage. The light will remain on until the autopilot is manually disengaged. According to Boeing, both the captain's and the first officer's control panels have flight mode annunciators to annunciate "A/P [autopilot] TRIM."

1.6.2 Design and Certification of MD-80 Series Airplanes

1.6.2.1 General

The original DC-9 was certified in 1965 under *Civil Aeronautics Regulations* (CAR) 4b, as revised through Amendment 12, March 30, 1962, as were other early models in the DC-9 series. The MD-80 was certified in 1980. More recent models of the DC-9, MD-80, MD-90, and 717 were certified under 14 CFR Part 25 and applicable amendments. However, systems that were similar to or that did not change significantly from the earlier DC-9 models, such as the longitudinal trim control system, were not required to be recertified under 14 CFR Part 25. Therefore, CAR 4b remained the certification basis for those parts of the MD-80, MD-90, and 717.

1.6.2.2 Longitudinal Trim Control System Certification

Pertinent sections of CAR 4b that were applicable to the original certification of the DC-9's longitudinal trim control system are outlined below.

CAR Section 4b.201(a), "Strength and Deformation," stated that structure "shall be capable of supporting limit loads without suffering detrimental permanent deformation."[40] CAR Section 4b.270, "Fatigue Evaluation of Flight Structure," stated the following:[41]

---

[40] Substantially similar requirements are currently contained in 14 CFR 25.305.

[41] Substantially similar requirements are currently contained in 14 CFR 25.571.

The strength, detail design, and fabrication of those portions of the airplane's flight structure in which fatigue may be critical shall be evaluated in accordance with the provisions of either paragraph (a) or (b) of this section.

(a) *Fatigue strength.* The structure shall be shown by analysis and/or tests to be capable of withstanding the repeated loads of variable magnitude expected in service....

(b) *Fail safe strength.* It shall be shown by analysis and/or tests that catastrophic failure or excessive structural deformation, which could adversely affect the flight characteristics of the airplane, are not probable after fatigue failure or obvious partial failure of a single principal structural element. After such failure, the remaining structure shall be capable of withstanding static loads corresponding with the flight loading condition specified [in CAR 4b]. These loads shall be multiplied by a factor of 1.15 unless the dynamic effects of failure under static load are otherwise taken into consideration.

CAR Section 4b.320(d), "Control Systems," stated the following:[42] "Each adjustable stabilizer must have a means to allow any adjustment necessary for continued safety of flight after the occurrence of any reasonably probable single failure of the actuating system."

CAR Section 4b.322, "Trim Controls and Systems," described the trim control systems requirements as follows:[43]

(a) Trim controls shall be designed to safe guard against inadvertent or abrupt operation....

(e) Trim devices shall be capable of continued normal operation in the event of failure of any one connecting or transmitting element of the primary flight control system.

CAR Section 4b.325, "Control System Stops," stated the following:[44]

(a) All control systems shall be provided with stops which positively limit the range of motion of the control surfaces.

(b) Control system stops shall be so located in the systems that wear, slackness, or take-up adjustments will not affect adversely the control characteristics of the airplane because of a change in the range of surface travel.

(c) Control system stops shall be capable of withstanding the loads corresponding with the design conditions for the control systems.

---

[42] Substantially similar requirements are currently contained in 14 CFR 25.671.

[43] Substantially similar requirements are currently contained in 14 CFR 25.677.

[44] Substantially similar requirements are currently contained in 14 CFR 25.675.

CAR Section 4b.606, "Equipment, Systems, and Installations," Subsection (b), "Hazards," stated, "All equipment, systems, and installations shall be designed to safeguard against hazards to the airplane in the event of their malfunctioning or failure." Similar requirements are currently contained in 14 CFR 25.1309. Specifically, 14 CFR 25.1309 requires that airplane systems and associated components be designed so that "the occurrence of any failure condition which would prevent the continued safe flight and landing of the airplane[45] is extremely improbable." Further, FAA AC 25.1309-1A, "System Design and Analysis," defines "extremely improbable" failure conditions as "those so unlikely that they are not anticipated to occur during the entire operational life of all airplanes of one type" and "having a probability on the order of 1×10^{-9} or less each flight hour, based on a flight of mean duration for the airplane type." AC 25.1309-1A specifies that in demonstrating compliance with 14 CFR 25.1309, "the failure of any single element, component, or connection during any one flight...should be assumed, regardless of its probability," and "such single failures should not prevent continued safe flight and landing, or significantly reduce the capability of the airplane or the ability of the crew to cope with the resulting failure condition."

According to Boeing, the longitudinal trim control system on the DC-9 was similar to the one used on earlier Douglas DC-8 (DC-8) airplanes. However, the DC-8 was equipped with a dual jackscrew design that provided structural and operational redundancy. According to Boeing, even though the DC-9 had a single jackscrew assembly, structural redundancy was accomplished by inserting a torque tube within the acme screw.

A Safety Board review of DC-9, MD-80/90, and 717 series airplane certification documents[46] indicated that the manufacturer provided material strength and loads information, stress calculations, and analyses to support assertions that the acme screw and torque tube were designed so that each of these two components could withstand tensile and compressive loads well beyond that which could be generated by aerodynamic forces acting on the horizontal stabilizer.[47] The documents also indicated that these assertions were based on the assumption of a new, intact acme screw and nut that met design specifications and that the acme screw and nut threads were intact and engaged to act as a load path.

Certification documents also included calculations and analyses to support the assertions that the jackscrew assembly was capable of carrying limit loads, with a margin

---

[45] A failure that would prevent continued safe flight and landing is referred to in Advisory Circular (AC) 25.1309-1A as a "catastrophic" failure condition.

[46] The Safety Board requested all available documentation addressing the certification of these airplane designs. Documentation provided by Boeing included the April 15, 1965, DC-9 Flight Controls System Fault Analysis Report (revised July 14, 1997); the 1964 DC-9 Control System Design Criteria; and the April 1998 MD-95-30 Control Systems Loads and Criteria Report. (The MD-95 was renamed the Boeing 717 when McDonnell Douglas and Boeing merged.) Internal design memorandums that addressed the horizontal stabilizer trim system were also reviewed.

[47] According to certification documents provided by Boeing, the DC-9/MD-80 series horizontal stabilizer acme screw was designed to carry an ultimate tension load of about 53,000 pounds. This load was calculated by assuming the highest aerodynamic loads that could be safely carried by the horizontal stabilizer structure in flight and then multiplying that load by 150 percent.

of safety, in either pure tension or compression, without failing, for each of the following scenarios:

- **A fractured acme screw.** According to the certification documents, this was accomplished by designing the inner torque tube with the capability to support limit operational tension and compression loads (in addition to the normal operational torsion loads) generated by the horizontal stabilizer.

- **A fractured torque tube.** According to the certification documents, this was accomplished by designing the jackscrew assembly so that if a torque tube fractured, trim operation would immediately cease. Under this failure scenario, the loads generated by the horizontal stabilizer would continue to be carried by the acme screw and nut, and pitch control would still be possible through the elevators.

- **The loss of 90 percent of acme screw and nut threads.** According to the certification documents, this was accomplished by designing the acme screw and nut with more threads than were structurally necessary.

- **The failure of one entire thread spiral in the acme screw or nut**. According to the certification documents, this was accomplished by incorporating two independent thread spirals along the acme screw's length. A Boeing Airplane Services structures engineering manager who testified at the Safety Board's public hearing on the flight 261 accident confirmed that "the reason behind having two independent threads is that if one of the threads were to fail, the other thread will carry the load."

A Safety Board review of the April 15, 1965, DC-9 Flight Controls System Fault Analysis Report (revised July 14, 1997) indicated that no contingency for the complete loss of the acme nut threads was incorporated into the design for the longitudinal trim control system. According to the fault analysis, the acme nut was designed with a softer material than the acme screw and its threads were designed to wear. Acme nut threads are made of an aluminum-bronze alloy and are about 0.15 inch thick at the minor diameter when new. The acme screw threads are made of case-hardened steel.

The DC-9 jackscrew assembly was originally designed for a service life of 30,000 flight hours and was not originally subject to periodic inspections for wear. In 1966, 1 year after the DC-9 went into service, the discovery of several assemblies with excessive wear resulted in the development and implementation of an on-wing end play check procedure to measure the gap between the acme screw and nut threads as an indicator of wear.[48] Thereafter, Douglas guidance specified that acme nut thread wear periodically be measured using an end play check procedure, and the acme nut was to be replaced when the specified end play measurement (0.040 inch) was exceeded.[49]

An FAA senior Aircraft Certification Office aerospace engineer testified at the public hearing that the jackscrew assembly was considered "primary structure because it

---

[48] End play is an indirect measurement of acme nut thread wear. For more information, see sections 1.6.3.3.1 and 1.6.3.3.2.

is the primary load-carrying load path from the horizontal [stabilizer] to the vertical [stabilizer], other than the pivot attaches at the rear spar." The FAA certification engineer also testified that it was not considered a principal structural element and added that "not all primary structure is a principal structural element." He defined a principal structural element, "in damage tolerance philosophy," as an element "whose failure, if it remained undetected, could lead to loss of the aircraft."

The FAA senior certification engineer further testified that the "jackscrew assembly and acme nut in particular…have to comply with the ultimate strength and limit load deflection criteria which provides for their strength." However, according to the FAA certification engineer, "because of the evaluation which indicates that the fatigue stresses are so low as to preclude the initiation or a propagation of a crack in the thread, they need not be considered under Part A or Part B of [CAR] 4b.270 [Fatigue Evaluation of Flight Structure]." According to the FAA certification engineer, who was referring to Boeing information, the "normal flight load…in the jackscrew to be transmitted from the screw through the nut to the vertical stabilizer is in the neighborhood of 4,000 or 5,000 pounds."

A Boeing structures engineering manager testified at the public hearing that "if the torque tube fails, the acme screw will carry the load, and if the acme screw fails, the torque tube will carry the load." The Boeing manager also stated that "we have no reported cases of fatigue failures or fatigue damage on this torque tube design for the four basic models, DC-9, MD-80, MD-90 and 717, and that is over 95 million accumulated flight hours, and…over 2,300 airplanes [were] delivered."

The FAA senior certification engineer testified at the public hearing that the jackscrew assembly was a "combination structural element and systems element…and [that] as such, the systems portion would fall under the systems requirements, and the structures portion would be required to address the structural requirements." The FAA certification engineer also indicated that the acme nut was not considered a system and, therefore, was not required to comply with airplane systems' certification requirements.

The Boeing structures engineering manager testified at the public hearing that "the acme nut thread is a wear component. Since the two independent threads in the acme nut are both subject to wear due to jackscrew rotation, we took this into account when we designed the system and defined the maintenance procedures to manage and inspect the wear.…When the acme nut is new, the strength of the threads is over 15 times in excess of design loads." The Boeing manager further testified that an acme screw and nut "unit worn to an end play reading of 0.040 of an inch remains 10 times stronger [than the maximum design loads]." He added, "Taking the end play reading to 0.080 of an inch, or twice the removal threshold, still leaves us with an installation five times stronger than the maximum design loads. [The] current in-service end play limit 0.040 of an inch is conservative from both the strength and the design standpoint. This assembly is

---

[49] Before the accident, there was no requirement to record or track end play measurements obtained during end play check procedures. After the accident, Airworthiness Directives (AD) 2000-03-51 and 2000-15-15 required that end play measurements be reported to Boeing. For more information about these ADs, see section 1.18.1.

recognized as a wear component which is addressed initially by its robust design and later by in-service maintenance action to control and monitor that wear."

The FAA senior certification engineer testified that wear is "not considered as a mode of failure for either a systems safety analysis or for structural considerations." The FAA certification engineer further stated that the "Douglas design of the acme nut and acme screw provided enough over-strength so that the regulatory requirements could be met with a significant amount of wear." He indicated that even if 14 CFR 25.1309 had been applicable to the entire jackscrew assembly, acme nut thread wear would not have been considered in the required systems safety analysis. He explained, "if you refer to a [section 25.1309] type safety analysis to try and determine a failure rate for a wear item, there are not, and there's really no such thing as a wear-critical item, never has been. The question of wear being a quantifiable element so that one could do a safety-type analysis for structure—it's not feasible. The data to do such an evaluation is not available. It doesn't exist."[50]

The FAA senior certification engineer stated that regulatory definitions of type design safety assume that the "safety of that type design is maintained…[because] the type certificate holder is required to recommend a maintenance program for [operators] such that the level of safety is maintained….We presume that those maintenance requirements are complied with." The FAA certification engineer added, "wear is considered as part of a design…and, as with any mechanism, maintenance is necessary." He concluded by stating, "I believe that the current design of the MD-80, DC-9, 717 and MD-90 pitch trim system demonstrates compliance to the applicable regulations and provides the level of safety associated with those certification requirements."

1.6.3 Maintenance Information

1.6.3.1 Alaska Airlines' MD-80 Maintenance Program

1.6.3.1.1 Maintenance Program Development Guidance

In 1968, an aviation industry team Maintenance Steering Group (MSG) developed a maintenance program development document for the then-new Boeing 747 (747). This guidance document contained decision logic and procedures for use in developing initial minimum maintenance and inspection requirements. Using the experience gained through the 747 project, in 1970, another MSG developed an updated guidance document, known as the MSG-2, which was designed to be applicable to all future newly type-certificated aircraft and/or powerplants, not just the 747. According to the MSG-2 guidance document, it was intended "to present a means for developing a maintenance program" that will be acceptable to the FAA, operators, and manufacturers.

In September 1980, through the combined efforts of the FAA, the Air Transport Association of America, U.S. and European aircraft and engine manufacturers, and U.S.

---

[50] The Boeing manager confirmed that "the condition of wear or wear-out was not included in the original DC-9 fault analysis."

and foreign airlines, a third guidance document, known as MSG-3, was issued. The MSG-3 guidance document, which incorporated data from more than 10 years of airline maintenance programs analysis and service experience, was developed, in part, to reduce confusion about how to interpret maintenance processes defined in the MSG-2 methodology, including hard-time, on-condition, condition-monitoring, and overhaul.[51] The MSG-3 task-oriented logic methodology used a "consequence of failure approach." For example, servicing and lubrication were included in the MSG-3 guidance document as part of the logic diagram to ensure that this category of task was considered each time an item was analyzed. After using MSG-3 analysis procedures on a number of new aircraft and powerplants in the first half of the 1980s, the airline industry decided that the experience should be used to improve the document for future applications, and, in 1988, Revision 1 to MSG-3 was issued; Revision 2 was issued in 1993.

The MSG-2 and MSG-3 guidance documents are accepted by the FAA and are used by Maintenance Review Boards (MRB) that are convened to develop reports containing guidance specific to an airplane type or model. MRB reports outline the recommended initial minimum scheduled maintenance and inspection requirements for a particular transport-category airplane and on-wing engine program. MRB reports are initially developed by an industry steering committee that is headed by an FAA chairperson. The steering committee directs working groups that include representatives from aircraft manufacturers and operators. The working groups are responsible for the development of initial minimum scheduled maintenance and inspection requirements for new or derivative aircraft to be used in the development of an operator's continuing airworthiness maintenance program. MRB reports are published and distributed by the manufacturer, but they contain FAA requirements. According to FAA AC 121-22A, the MRB report is "a means, in part, of developing instructions for continuing airworthiness, as required by [14 CFR section] 25.1529." The MRB report is submitted to the FAA for approval as part of these instructions.

On the basis of the guidance in the applicable MRB report, manufacturers issue on-aircraft maintenance planning (OAMP) documents and generic task cards outlining specific maintenance tasks. The tasks and intervals in an OAMP document are typically the same as those contained in the associated MRB report. Two MRB reports and associated OAMP documents are currently used as the basis for the maintenance of the MD-80: those derived from MSG-2, issued in 1993, and those derived from MSG-3, issued in 1996. Carriers can use the guidance in the MRB report and OAMP documents to

---

[51] Hard-time is a maintenance process that requires an item to be removed from service or overhauled at or before a previously specified time. On-condition is a maintenance process that requires periodic inspections or checks of a unit against an appropriate physical standard to determine whether it can continue in service until the next scheduled check and to remove the unit from service before failure occurs during normal operations. Condition-monitoring is a maintenance process for items that do not have hard-time or on-condition as the primary maintenance process and is accomplished through continuous data collection and analysis of components or systems. Analysis of failures or other indications of deterioration are used to evaluate the continuing airworthiness of the airplane. Overhaul is the disassembly, cleaning, inspection, repair, and testing of a component to the extent necessary to ensure (substantiated by service experience and accepted practices) that it is in satisfactory condition to operate for a given period. It includes the replacement, repair, adjustment, or refinishing of such parts as required, which, if not properly accomplished, would adversely affect the airworthiness of the airplane.

develop their maintenance programs.[52] Carriers with maintenance programs based on the MSG-2 MRB report are permitted to use the MSG-3 MRB report and its listed intervals to adjust existing intervals in coordination with the airline's FAA principal maintenance inspector (PMI).[53]

1.6.3.1.2 Alaska Airlines' Continuous Airworthiness Maintenance Program

Alaska Airlines' continuous airworthiness maintenance program for its MD-80 fleet received initial FAA approval in March 1985 and was based on the FAA-accepted MRB report for DC-9/MD-80 airplanes derived from the MSG-2 maintenance program development document. The MRB report and the resulting OAMP document recommended C-check intervals of 3,500 flight hours or 15 months, whichever came first. Alaska Airlines' initial C-check interval was 2,500 flight hours. In 1988, it was extended to every 13 months (which, based on the average airplane utilization rate at Alaska Airlines at the time, was about 3,200 flight hours). In 1996, the C-check interval was extended to 15 months (which, based on the average airplane utilization rate at Alaska Airlines at the time, was about 4,775 flight hours).

In addition to the 15-month C-check interval, Alaska Airlines' General Maintenance Manual (GMM), dated February 5, 1998, revealed that Alaska Airlines had the following maintenance inspections and intervals for its MD-80 fleet at the time of the accident:

- Walk-around check—accomplished by Alaska Airlines maintenance personnel after each arrival.

- Service check—accomplished when the airplane remains on the ground for 8 or more hours at a maintenance facility capable of conducting at least A check maintenance.

- A check—accomplished at intervals not to exceed 250 flight hours.

1.6.3.1.3 Alaska Airlines' Reliability Analysis Program

CFR Section 121.373(a), "Continuing Analysis and Surveillance," requires operators to establish and maintain a system for the continuing analysis and surveillance of the performance and effectiveness of their maintenance and inspection programs and for the correction of any deficiency found in those programs. According to FAA Order 8300.10, chapter 37, continuing analysis and surveillance programs should have, as a minimum, procedures for monitoring mechanical performance, which should be done

---

[52] FAA AC 121-22A states that an MRB report "contains the initial minimum scheduled maintenance/inspection requirements for a particular…aircraft…[it should] not to be confused with, or thought of, as a maintenance program. After approval by the FAA, the requirements become a base or framework around which each air carrier develops its own individual maintenance program."

[53] Alaska Airlines used the MSG-2 MRB report as the basis for its maintenance program; however, it used the MSG-3 MRB report to extend some maintenance task intervals by replacing them with the newer intervals specified in the MSG-3 MRB report. The FAA's MD-80 MRB Chairman confirmed that existing MSG-2 operators, such as Alaska Airlines, could use intervals from MSG-3. Alaska Airlines' manager of reliability and maintenance programs testified that the airline had done so in some areas.

through emergency response, day-to-day and long-term monitoring, and auditing functions. The handbook states that "some operators with approved reliability programs use the reliability program to fulfill the monitoring mechanical performance functions requirement of its [continuing analysis and surveillance] program." Typical sources of data collection for reliability programs are unscheduled removals, mechanical delays and cancellations, pilot reports, sampling inspections, shop findings, functional checks, bench checks, service difficulty reports, and mechanical interruption summaries. The objective of data analysis is to recognize the need for corrective action, establish what corrective action is needed, and determine the effectiveness of that action. Other indicators of component reliability used to detect unfavorable trends are operational incident reports and direct information from maintenance control, component shops, or repair control about component discrepancies.

Alaska Airlines' reliability analysis program was authorized on April 3, 1995, as part of the airline's FAA-approved operations specifications. According to reliability analysis program guidelines, Alaska Airlines managed its maintenance program by implementing decisions based on continuing analysis of operational data. All scheduled maintenance/inspection tasks, checks, and intervals were included in the reliability analysis program. Reliability analysis program findings could result in changes, additions, or deletions of maintenance tasks, intervals, or operating limits.

Alaska Airlines has made changes to its reliability analysis program since the flight 261 accident. This discussion describes Alaska Airlines' reliability analysis program up until the time of the accident. The Reliability Analysis Program Control Board, which reported to Alaska Airlines' vice president of maintenance and engineering, was responsible for the overall program.[54] The Reliability Analysis Program Control Board was authorized to determine what course of remedial action was necessary and to order implementation of its decisions. The Reliability Department, which reported to the Reliability Analysis Program Control Board, administered the program and was responsible for reviewing and analyzing collected data, issuing alert notices, and providing notification of reliability issues.

The Reliability Analysis Program Control Board also included a smaller Reliability Action Board that addressed problems considered critical or urgent and that had to be considered before the next Reliability Analysis Program Control Board meeting.[55] Reliability Action Board decisions that required changes to the activities controlled by the reliability analysis program required the preparation and approval of a Reliability Analysis Program Control Board directive.

---

[54] Alaska Airlines Reliability Analysis Program Control Board, which met once per month, comprised the airline's reliability manager (who served as board chairman), the director of base maintenance, the director of line maintenance, the director of quality control, the director of engineering, the manager of maintenance control/technical services, the manager of maintenance programs and publications, and the director of maintenance planning and production control.

[55] A Reliability Action Board meeting did not require the attendance of the full Reliability Analysis Program Control Board but had to be attended by the Reliability Analysis Program Control Board Chairman (or designee) and other members, as appropriate to the issue being discussed.

In addition, a company MRB at Alaska Airlines met biweekly to review and approve changes to the maintenance program.[56] The MRB also reviewed ME-01 change requests[57] that required MRB approval. Under Alaska Airlines' MRB program, the manager of maintenance programs and publications reviewed ME-01 change requests to determine whether the change required MRB approval. If the ME-01 did not require MRB/Reliability Analysis Program Control Board approval, the request was routed to management personnel in the following order: reliability, director of engineering, director of line maintenance, director of base maintenance, manager of maintenance control, director of quality control, director of maintenance planning and production control. The MRB/Reliability Analysis Program Control Board voting member representing the department that requested the change reviewed the ME-01 and then decided whether to sign it. If signed, the ME-01 is forwarded to the manager of the maintenance programs and technical publications department for processing.

MRB approval was required for any change to Alaska Airlines' GMM. Reliability analysis program approval was required for any maintenance program change, some of which required FAA approval. According to Alaska Airlines' reliability analysis program, maintenance program changes that required FAA approval included the following:

- Addition or deletion of an aircraft manufacturer's MSG-2 OAMP recommended maintenance significant item (MSI),[58] or an aircraft manufacturer's MSG-3 OAMP document recommended systems maintenance, structural, or zonal task.[59]

- Addition or deletion of an MSG-2 MRB recommended MSI or an MSG-3 MRB recommended systems maintenance, structural, or zonal task.

- Escalation of hard-time component overhaul intervals, periodic maintenance check intervals, or structural inspection intervals to greater than 10 percent of the present interval.

- Escalation of an individual maintenance task interval (currently performed at or in excess of the interval recommended in the applicable aircraft's OAMP document) to greater than 10 percent of the current interval.

Changes that did not require FAA approval included the following:

---

[56] Alaska Airlines' company MRB members were the same as those who served on the Reliability Analysis Program Control Board.

[57] The ME-01 form is used by Alaska Airlines to process requests and recommendations to change existing maintenance program procedures, intervals, and documents (task cards, maintenance manuals, and policy and procedures manuals), and other documents under the control of the airline's engineering and quality departments. The ME-01 is not required for typographical errors, P/N changes, clarification of existing procedures, changes to torques, clearances and limits required to reflect manual revisions, and changes generated by previously authorized airplane modifications or actions to meet AD or *Federal Aviation Regulations* (FAR) requirements.

[58] An MSI is an item identified by the manufacturer, the failure of which is undetectable during operations and could affect safety (on ground or in flight) or could have significant operational economic impact.

[59] Zonal tasks are tasks associated with specific areas of the airplane, such as the wing and fuselage.

- Escalation of hard-time component overhaul intervals, periodic maintenance check intervals, or structural inspection intervals that did not exceed 10 percent of the current interval. (Although FAA approval is not required for these escalations, data justifying the change must be available to the FAA on request.)

- Escalation of an individual maintenance task interval (which is not listed on the airplane's MRB) up to the interval recommended in the applicable airplane's OAMP document, regardless of the increase of the current interval.

- Periodic maintenance intervals of MSIs identified as condition-monitored items, which have a periodic maintenance activity or task at some determined interval.

Items that cannot be changed or that required approval beyond the local FAA Flight Standards District Office (FSDO) included the following:

- Those tasks listed as certification maintenance requirements (CMR),[60] airworthiness limitation items,[61] or fixed maintenance intervals unless approval is granted from the FAA Aircraft Certification Office.

- Life limits mandated by the manufacturer.

- Time limits mandated by an AD, except in accordance with the AD.

- Time limits mandated by an aircraft's type certificate data sheet.

1.6.3.2 Jackscrew Assembly Lubrication Procedures

The Alaska Airlines maintenance task card for elevator and horizontal stabilizer lubrication (No. 28312000) specified the following steps for lubricating the jackscrew assembly:[62]

A. Open access doors 6307, 6308, 6306 and 6309.

B. Lube per the following

---

[60] According to AC 25-19, "Certification Maintenance Requirements," dated November 28, 1994, "a CMR is a required periodic task, established during the design certification of the airplane as an operating limitation of the type certificate...[and is] intended to detect safety-significant latent failures that would, in combination with one or more other specific failures or events, result in a hazardous or catastrophic failure condition." According to the AC, "CMRs are designed to verify that a certain failure has or has not occurred, and does not provide any preventative maintenance function." (Emphasis in original.)

[61] Airworthiness limitation items are documents that contain information that defines mandatory replacement times for safe-life structure and inspection requirements for principal structural elements and provide specific nondestructive inspection techniques and procedures for each principal structural element.

[62] Illustrations were attached to the task card to show the areas to be lubricated.

ITEM NO.	ITEM DESCRIPTION	NO. OF FITTINGS OR AREAS
1.	MAIN GEAR BOX	1
	Check oil level and fill to fill plug with approved jet engine oil. Oil capacity is approximately 0.6 pint.	
2.	ACTUATOR ASSEMBLY	1
3.	JACKSCREW	0
	Apply light coat of grease to threads, then operate mechanism through full range of travel to distribute lubricant over length of jackscrew.	
4.	SCREW GIMBAL	3
5.	HORIZONTAL STABILIZER HINGE	2
6.	JACKSCREW STOP	0
	Fill cavity between screw and stop with Parker-O-Lube.	

C. Close doors 6307, 6308, 6306 and 6309.

General lubrication guidance contained in Boeing's Maintenance Manual states that, when lubricating vented bearings, "force grease into fittings until all old grease is extruded." According to a February 25, 2002, letter from Boeing to the Safety Board, this lubrication task requires about 4.5 person hours to accomplish, including removal and closure of the access panels.

During the investigation, Safety Board investigators conducted interviews with Alaska Airlines mechanics who had lubricated the accident airplane's elevators and horizontal stabilizer components to determine how they understood and practiced the lubrication procedures. The Alaska Airlines mechanic who was responsible for lubricating the accident airplane's elevators and horizontal stabilizer components during a C check at the airline's maintenance facility at Oakland International Airport (OAK), Oakland, California, in September 1997[63] told Safety Board investigators that lubrication of the elevators and horizontal stabilizer components "takes a couple of hours." The mechanic stated that if the lubrication of the elevators and horizontal stabilizer components is accomplished during a C check, the task is accomplished by a mechanic standing on a tail-stand platform in the hangar. He stated that he would apply grease to acme nut grease fittings until "the grease starts to come out…the sides where you're greasing, where it starts to bulge out." The mechanic stated that he would also use a "paint brush" to apply grease to the acme screw threads. He added, "When I do it, I do the brush, too, but I don't feel it's enough. I always do an over-kill. I always like to put more on…[to] put a big glob on my hand and make sure it's on. More's better. It's not going to hurt anything."

The Alaska Airlines mechanic who was responsible for performing the accident airplane's last elevator and horizontal stabilizer components lubrication at SFO in September 1999 told Safety Board investigators that he would know if the grease fitting for the acme nut was clogged "because you'd feel it in the [grease] gun as you try to put it in. If it wasn't going in, you could feel it." When asked how he determined whether the lubrication was being accomplished properly and when to stop pumping the grease gun, the mechanic responded, "I don't." When asked whether he would be able to see grease coming out of the top of the acme nut during lubrication, the mechanic responded, "You know, I can't remember looking to see if there was." He added that lubrications at SFO were accomplished most often during the evening, outside of the hangar (even if it was raining), and in the basket of a lift-truck. He added that he used a battery-operated head lamp during night lubrication tasks. He stated that the lubrication task took "roughly…probably an hour" to accomplish. It was not entirely clear from his testimony whether he was including removal of the access panels in his estimate. When asked whether his 1-hour estimate included gaining access to the area, he replied, "No, that would probably take a little—well, you've got probably a dozen screws to take out of the one panel, so that's—I wouldn't think any more than an hour." The questioner then stated, "including access?" and the mechanic replied, "Yeah." The mechanic also stated that he applied grease to the acme screw with a brush, adding, "You basically take out a little bit of grease with you, or you pump it out of the grease gun onto a paint brush and just paint a real light coat on there."[64]

1.6.3.2.1 Jackscrew Assembly Lubrication Intervals

1.6.3.2.1.1 Manufacturer-Recommended Lubrication Intervals

Original DC-9 certification documents, including Douglas Process Standard 3.17-49 (issued August 1, 1964), specified a lubrication interval for the jackscrew assembly of 300 to 350 flight hours. The manufacturer's initial OAMP document for the DC-9 and the MD-80 specified lubrication intervals of 600 to 900 flight hours. In 1996, McDonnell Douglas issued another OAMP document, which was developed to reflect the guidance and philosophy in the MSG-3 MSG-3 MRB report for the MD-80. The MSG-3 MRB called for lubrication of the jackscrew assembly at C-check

---

[63] There were three scheduled lubrications after the September 1997 C check. The C check performed in September 1997 was the last C check accomplished on the accident airplane that required an end play check of the acme screw and nut. For more information about the accident airplane's maintenance history, see section 1.6.3.5.

[64] For information about Safety Board observations of jackscrew assembly lubrications performed by maintenance personnel from two MD-80 operators, see section 1.18.6.

intervals (every 3,600 flight hours or 15 months, whichever comes first). The MSG-3 OAMP also called for lubrication of the jackscrew assembly at C-check intervals (every 3,600 flight hours).[65]

Testimony at the Safety Board's public hearing and Boeing documents indicated that Douglas' earlier recommended lubrication interval of 600 to 900 flight hours was not considered during the MSG-3 decision-making process to extend the recommended interval. Further, Boeing design engineers were not consulted about nor aware of the extended lubrication interval specified in the MSG-3 documents. The FAA's MD-80 MRB chairman testified at the public hearing that the longer interval in the MSG-3 documents was supported by reliability data from the carriers and the manufacturer. He explained that the MSG-3 MD-80 MRB considered the jackscrew lubrication task change as part of a larger C-check "package" and that the extension of C-check intervals in the MSG-3 MD-80 MRB did not involve a task-by-task analysis of each task that would be affected by the changed interval.

1.6.3.2.1.2 Alaska Airlines' Lubrication Intervals

Alaska Airlines maintenance records chronicled the following changes in lubrication intervals from 1985 to April 2000:

- In March 1985, the B-check interval was 350 flight hours, and the jackscrew assembly lubrication was to be accomplished at every other B check, or every 700 flight hours.

- In March 1987, the B-check interval was increased to 500 flight hours, and the jackscrew assembly lubrication was changed to be accomplished at every B check, or every 500 flight hours.

- In July 1988, B checks were eliminated from Alaska Airlines' MD-80 maintenance program, and the tasks accomplished during B checks were incorporated into the A and C checks. A checks were to be accomplished every 125 flight hours. The jackscrew assembly lubrication was to be accomplished at every eighth A check, or every 1,000 flight hours.

- In February 1991, A-check intervals were increased to 150 flight hours. Jackscrew assembly lubrication was still to be accomplished at every eighth A check, or every 1,200 flight hours.

- In December 1994, the A-check interval was increased to 200 flight hours. Jackscrew assembly lubrication was still to be accomplished at every eighth A check, or every 1,600 flight hours.

- In July 1996, the jackscrew assembly lubrication task was removed from the A check and placed on a time-controlled task card with a maximum interval of 8 months.[66] There was no accompanying flight-hour limit; however, based on airplane utilization rates at that time, 8 months was about 2,550 flight hours.

---

[65] The FAA's MD-80 MRB Chairman testified at the public hearing that he assumed the absence of any calendar-time interval in the MSG-3 OAMP document was a typographical error.

Jackscrew assembly lubrication was also to be performed at each C check using time-controlled task card No. 28312000. According to the FAA PMI for Alaska Airlines, who reviewed and accepted the 1996 lubrication interval extension, the extended MSG-3 interval (3,600 flight hours or 15 months, whichever comes first) was presented as justification for the interval increase.

- In October 1996, jackscrew assembly lubrication was combined with the lubrication of the elevators and elevator tabs. The interval remained at 8 months, or about 2,550 flight hours, and this was the lubrication interval in effect at the time of the accident.

- On April 6, 2000, after the accident, the jackscrew assembly lubrication interval was changed from 8 months to 650 flight hours, in accordance with ADs 2000-03-51 and 2000-15-15.[67] Table 2 shows a comparison of manufacturer-recommended jackscrew assembly lubrication intervals with Alaska Airlines' intervals.

Table 2. Comparison of manufacturer-recommended jackscrew assembly lubrication intervals with Alaska Airlines' intervals.

MSG-2 MRB	MSG-2 OAMP	MSG-3 MRB	MSG-3 OAMP
Not included in logic diagram	600 to 900 flight hours	C check (3,600 flight hours or 15 months, whichever comes first)	C check (3,600 flight hours)

Alaska Airlines 1985	Alaska Airlines 1987	Alaska Airlines 1988	Alaska Airlines 1991	Alaska Airlines 1994	Alaska Airlines 1996 to April 2000	Alaska Airlines April 2000 to Present[a]
Every other B check (700 flight hours)	B check (500 flight hours)	Every eighth A check (1,000 flight hours)	Every eighth A check (1,200 flight hours)	Every eighth A check (1,600 flight hours)	Time-controlled task card— 8 months maximum (About 2,550 flight hours)	650 flight hours

a. All carriers currently meet this requirement.

---

[66] A time-controlled task card is a stand-alone maintenance task card that is tracked individually by the operator.

[67] For more information about ADs 2000-03-51 and 2000-15-15, see section 1.18.1.

1.6.3.2.2 Chronology of Grease Type Changes

Boeing's Airplane Maintenance Manual for DC-9, MD-80/90, and 717 series airplanes specifies the use of MIL-G-81322[68] grease to lubricate the jackscrew assembly. Initially, Alaska Airlines used Mobilgrease 28, which meets the requirements of MIL-G-81322. On January 16, 1996, Alaska Airlines asked its McDonnell Douglas field representative if Aeroshell 33 could be used in place of Mobilgrease 28 on suitable areas of its MD-80 fleet airplanes, including the jackscrew assembly. Aeroshell 33 was a relatively new type of grease that was developed to meet Boeing Material Specifications 3-33 and subsequently qualified under MIL-G-23827.[69]

According to Alaska Airlines, this change was part of an effort to standardize and reduce the number of greases used by the airline in its mixed fleet of 737s and MD-80s. On February 23, 1996, McDonnell Douglas engineers replied that laboratory testing would be required to determine whether Aeroshell 33 could be approved for use on Douglas airplanes. Alaska Airlines was informed on January 24, 1997, that laboratory testing of Aeroshell had begun and that the testing could take about a year.

On June 19, 1997, Douglas informed its Alaska Airlines field service representative that it was pursuing the possibility of a "no technical objection" for Alaska Airlines to use Aeroshell 33 on Douglas airplanes. However, the message stated, "this is not a sure thing, considering the liability of providing consent for the use of such an important substance…before the substance has been fully evaluated." The message stated that another possibility was an in-service evaluation of the grease by Alaska Airlines. On June 23, 1997, the Douglas field service representative sent a message to Douglas headquarters noting that Alaska Airlines welcomed the offer to conduct an in-service evaluation of Aeroshell 33 and was awaiting guidelines regarding its use.

On July 23, 1997, Alaska Airlines initiated an internal form titled, "Maintenance Programs/Technical Publications Change Request" (ME-01 No. 97-002974), to revise its maintenance lubrication task cards by replacing Mobilgrease 28 with Aeroshell 33 for use on flight controls, including the jackscrew assembly, doors, and landing gear (except wheel bearings), on MD-80 series airplanes in its fleet. This form was signed by the initiator of the request, the manager of maintenance programs and technical publications (dated September 17, 1997), and the director of engineering (dated July 25, 1997). Signature blocks for the director of base maintenance and the director of maintenance planning and production control were blank but struck out by hand. The "special action

---

[68] The "MIL" specification is a U.S. military material specification that defines the requirements for a material. Requirements that are defined for grease MIL specification categories include, in part, viscosity, temperature range, and corrosion protection. A part's function and exposure to such factors as loads, friction, extreme temperatures, and water and other contaminants determine which grease MIL specification category is appropriate. In 1999, the MIL-G nomenclature (in which G stands for grease) was replaced with MIL-PRF (in which PRF stands for performance); no change in specifications was made. Several products from different manufacturers may meet a specific MIL specification. After testing and qualification, these greases are listed on the Qualified Products List for each MIL specification. For example, greases that meet the specifications for MIL-G-81322 include Mobilgrease 28, Aeroshell Grease 22, and Royco Grease 22.

[69] MIL-PRF-23827 superseded MIL-G-23827 in 1998, with no change in specifications.

required"[70] section was blank, as was the signature block at the bottom of the form to indicate "Change Request accomplished." At the Safety Board's public hearing, Alaska Airlines former director of engineering (who at the time that the grease change was initiated was Alaska Airlines manager of maintenance programs and technical publications) and the current director of engineering indicated that Reliability Analysis Program Control Board approval was required to implement the change. However, neither individual could explain how the change was implemented without the required signatures or Reliability Analysis Program Control Board approval.

On September 26, 1997, McDonnell Douglas informed Alaska Airlines in Douglas Communication SVC-SEA-0122/MRL that it had "no technical objection to the use of [Aeroshell 33] grease in place of MIL-G-81322 grease on Alaska Airlines MD-80 aircraft with one known restriction. Aeroshell 33 grease may not be used in areas subjected to temperatures in excess of 250° F including landing gear wheel bearings." The McDonnell Douglas communication added the following:

> It should also be noted that the initial results of laboratory testing that compared Aeroshell 33 grease with Mobile grease 28 (MIL-G-81322) indicated that Aeroshell 33 grease is somewhat less resistant to water wash-out than Mobile grease 28. It is not known whether the difference in water wash-out resistance noted in the laboratory will result in any notable difference of the in-service grease performance. However, the potential exists that the required frequency of lubrication could be affected in areas of the aircraft exposed to outside ambient conditions or aircraft washing and cleaning.

> This no technical objection is provided prior to the completion of a Douglas study intended to determine the acceptability of Aeroshell 33 grease for use in Douglas-built aircraft.[71] As such, Douglas cannot yet verify the performance of this grease. It will be the responsibility of Alaska Airlines to monitor the areas where Aeroshell 33 grease is used for any adverse reactions. Further, it will be the responsibility of Alaska Airlines to obtain any FAA approval required by their principal maintenance inspector for the use of Aeroshell 33 grease in their MD-80 aircraft.

The McDonnell Douglas communication concluded with the following:

> Because the Douglas study of Aeroshell 33 grease will be based on laboratory tests rather than an in-service test, any data we can obtain from in-service aircraft would be valuable to help confirm or refute the laboratory results. Any information Alaska Airlines is able to provide to Douglas, positive or negative, regarding the in-service performance of Aeroshell 33 grease on the MD-80 will be greatly appreciated.

---

[70] The possible choices in this section, each preceded by a block to check off, were "MRB action required," "[Reliability Analysis Program] Control Board action required," or "Routine."

[71] According to Boeing, McDonnell Douglas never completed this testing.

On December 18, 1997, Alaska Airlines task card No. 24312000 for lubrication of the horizontal stabilizer components was revised (based on ME-01 No. 97-002974) to reflect the change from Mobilgrease 28 to Aeroshell 33.[72] On January 6, 1998, task card No. 28312000 for lubrication of the elevators and horizontal stabilizer components was also revised to reflect the change from Mobilgrease 28 to Aeroshell 33.[73]

According to Alaska Airlines records, it sent the monthly maintenance task card audit report for December 1997, which included all task card changes since its last report, including revision dates, to the FAA. Although Alaska Airlines notified the FAA of the change in the grease type specified in its maintenance instructions, as required, it did not provide any substantiating justification at that time nor was such action required. The FAA PMI for Alaska Airlines at the time of the lubrication change stated during postaccident interviews that such changes could be made without prior FAA approval under provisions of the airline's FAA-approved maintenance program.[74]

As a result of the flight 261 accident, in March 2000, the FAA requested documentation from Alaska Airlines that supported the grease change to lubricate the jackscrew assembly. Alaska Airlines subsequently submitted the requested documentation.[75] In an April 5, 2000, response letter, the FAA informed Alaska Airlines that the substantiating documents it had submitted did not support the change, and it disapproved the use of Aeroshell 33 as a substitute for Mobilgrease 28. After receiving this disapproval notification, Alaska Airlines modified its maintenance procedures to again specify the use of Mobilgrease 28 to lubricate the jackscrew assembly.

1.6.3.2.3 July 19, 2002, Alaska Airlines Maintenance Information Letter

On July 19, 2002, Alaska Airlines issued Maintenance Information Letter No. 05-00-02-21, "MD-80 Rudder Trim Tab Hinge Support Bearings Lubrication,"[76]

---

[72] According to Alaska Airlines records, 34 additional lubrication task cards, including lubrication tasks for flaps, slats, spoilers, ailerons, and landing gear components, were revised in December 1997. These task cards also reflected the change to Aeroshell 33.

[73] The GMM was not modified to reflect the grease change to Aeroshell 33; it still specified the use of Mobilgrease 28 to lubricate the elevators and horizontal stabilizer components.

[74] According to the PMI, changes to "accepted [maintenance] manuals" used as part of an overall FAA-approved maintenance program can be made and then submitted to the FAA as part of a routine change notification process. The PMI stated, "in accepted manuals, they go ahead and make the change, publish it, and send you a copy and you read it. And if you have any objection to it, [you] notify them in writing that you have objections. If you don't tell them, then it's accepted." Referring to the Alaska Airlines lubrication change and the task card change notification in 1997, the PMI stated, "I don't know that anybody caught that or noticed it or bought off on it or looked into it at all."

[75] The copies provided to the Safety Board of these documents included a trade magazine article on Aeroshell 33, excerpts of Boeing 737 and McDonnell Douglas MD-80 maintenance manuals, Boeing service letters, internal correspondence and messages between Alaska Airlines and Boeing, and existing specifications on MIL-G-81322 and MIL-G-23827 greases. None of the documents provided any information or performance data specifically applicable to the use of Aeroshell 33 on McDonnell Douglas airplanes or these airplanes' horizontal stabilizer trim systems.

[76] Maintenance Information Letter No. 05-00-02-21 is valid for 1 year, expiring on July 19, 2003, or when rescinded.

because of "findings [that] have revealed that the MD-80 rudder trim tab support bearings may not always be receiving adequate lubrication." Alaska Airlines maintenance task card No. 24612000 specifies lubrication of each of the four bearings that support the MD-80 rudder trim tab at every C check. However, the letter stated that "shop inspection of tabs from two aircraft...only recently out of C-check indicated extreme bearing wear, with little or no evidence of new grease" and that "one bearing fell apart upon removal." The letter stated that the solution was adherence to the lubrication instruction in task card No. 24612000. In an October 3, 2002, letter to the Safety Board, Alaska Airlines indicated that the "[maintenance information letter] was incorrect when it described these aircraft as 'only recently out of C-check.'" Alaska Airlines stated that one of the airplanes had last been in for a C check about 6 months before the discovery of the worn bearing and that the other airplane had last been in for a C check about 15 months before the discovery of the worn bearing. In an October 24, 2002, facsimile to the Safety Board, Alaska Airlines stated further that as a result of additional analysis, the rudder trim tab lubrication interval had been reduced "from the current 'C' check (approximately 4,350 flight hours) to the Boeing (McDonnell Douglas) MSG-2 On Aircraft Maintenance Planning...'Low' end interval of 1,200 flight hours."

1.6.3.3 Procedures for Monitoring Acme Nut Thread Wear

1.6.3.3.1 Defining and Calculating Wear Rate

The only portion of the jackscrew assembly that wears significantly is the aluminum-bronze acme nut. Therefore, the terms "wear" and "wear rate" refer to the wear of the acme nut.

As will be discussed in section 1.6.3.3.2, acme nut thread wear can be inferred by measuring the amount of movement, or end play, between the acme nut and screw threads. In other words, end play is an indirect measurement of acme nut wear. End play measurements are typically performed either on-wing or after removal from the airplane during a bench check. The Safety Board conducted a statistical analysis of end play data that concluded that the reliability and validity of the on-wing end play check is very low and is susceptible to measurement error.[77] Because the bench end play check is conducted in a more controlled environment, it is presumably less susceptible to measurement error; however, the reliability and validity of the bench end play check was not statistically evaluated.

Unless otherwise specified, the term "wear rate" used in this report refers to an approximate wear rate determined by comparing end play measurements from the beginning and the end of an elapsed time. The amount of the change in end play over that period of time is used to calculate the approximate rate of wear. (Wear rate calculations are generally converted so that they can be expressed as an average wear rate per 1,000 flight hours.) The wear rate at any given point during the elapsed time is not known; it may be higher or lower than the approximated wear rate.

---

[77] For more information about the Safety Board's end play data study, see section 1.18.2.

Factual Information

1.6.3.3.2 Development of End Play Check Procedures

The DC-9 jackscrew assembly was originally certified with an expected service life of 30,000 hours. According to Douglas documents,[78] after 30,000 hours, the gap between the acme screw and nut threads was expected to be 0.0265 inch. Specifications for newly manufactured jackscrew assemblies stipulate that this gap be between 0.003 to 0.010 inch. No regular inspections to monitor acme nut thread wear were recommended at the time the DC-9 was certified. However, when the DC-9 entered service in 1965, Douglas initiated a program, which closely monitored a sample of DC-9 airplanes. As part of this program, an end play check procedure was used to monitor wear by measuring the gap, or end play, between the acme screw and nut threads.[79] This procedure was conducted during bench checks to determine whether the wear rate of the acme nut threads was comparable to the DC-8 wear rate of 0.001 inch per 1,000 hours.[80]

As a result of excessive wear rates discovered through this sampling program, Douglas issued Douglas Design Memorandum C1-250-DES-DC9-683 in February 1967. The memorandum stated that DC-9 operators had reported six jackscrew assemblies that indicated "a wear rate considerably in excess of that predicted"[81] and provided the following information about the flight hours and end play measurements from the six jackscrew assemblies: (The Safety Board calculated the wear rate per 1,000 hours, taking into account the end play measurement of the jackscrew assembly when new. If the end play measurement when new was known, that value was used. Otherwise, a presumed end play measurement of 0.007 inch was used.)

Airplane No.	Flight Hours	End Play (inches)	Wear Rate (inches per 1,000 hours)
1	4,086	0.030	0.0056
2	3,300	0.026	0.0058
3	3,322	0.026	0.0057
4	3,390	0.029	0.0065
5	3,400	0.028	0.0062
6	2,200	0.023	0.0073

---

[78] The documents were All Operators Letter (AOL) 9-48, dated November 4, 1966, and Douglas Design Memorandum C1-250-DES-DC9-683, dated February 28, 1967.

[79] The sampling program ended in 1970. As will be discussed, by this time, Douglas had determined that wear rates for the DC-9 jackscrew assembly made with new materials and manufacturing procedures were 0.001 inch per 1,000 hours or lower.

[80] A 1967 Douglas test report on DC-8 jackscrew assemblies indicated an average wear rate of not greater than 0.001 inch per 1,000 hours.

[81] The design memorandum also noted, "there is evidence from screws in service, which have been returned to [Douglas], that lubrication practices as defined in the Maintenance Manual have not been followed. Operators should be advised of the importance of relubrication as specified."

As a result of these operator reports, Douglas conducted a series of full-scale tests, which involved operating several jackscrew assemblies under varying conditions in a test fixture. The Douglas engineering test report stated that the purposes of the tests were "to simulate a rate of wear in the laboratory that is comparable to that being experienced in service on the DC-9 horizontal stabilizer screw and nut [and to] determine the right material of the jackscrew end nut, and/or lubrication, which would generate a minimum of wear to provide a service life of 30,000 flight hours." As a result of these tests, Douglas established the end play measurement limit of 0.040 inch after determining that this limit would not create aerodynamic flutter or other structural problems.[82] Douglas also changed some of the materials specifications and manufacturing processes for the acme screw (increased heat treating and nitriding) to reduce wear. Douglas conducted laboratory tests of the DC-9 acme screws made with the newly specified materials and manufacturing processes, which revealed that the screws had an average wear rate of about 0.001 inch per 1,000 hours.

To allow operators to monitor acme nut wear without removing the jackscrew assembly from the airplane, Douglas developed the on-wing end play check procedure. The procedure calls for pulling down on the horizontal stabilizer by applying a specified amount of torque to a tool known as a restraining fixture to change the load on the acme screw from tension to compression.[83] The load reversal allows movement, or end play, between the acme screw and nut threads. This play is measured using a dial indicator, which is mounted on the lower rotational stop (fixed to the screw) with the movable plunger set against the lower surface of the acme nut, that measures the vertical movement of the acme screw when the load is applied. The amount of end play is used as an indication of the amount of acme nut thread wear (and acme screw thread wear, if there is any). The procedure calls for the restraining fixture load to be applied and removed several times until consistent measurements are achieved. Figure 8 shows how the restraining fixture should be placed on the horizontal stabilizer during the end play check procedure. Figure 9 shows a view of a typical dial indicator set up for end play checks.

---

[82] In AOL 9-48A, Douglas stated that jackscrew assemblies could remain in service as long as the end play measurement remained within tolerances (between 0.003 and 0.040 inch). End play checks that indicated measurements within the specified limits were not required to be recorded before the issuance of AD 2000-15-15 in July 2000.

[83] For more information about restraining fixtures, see section 1.6.3.3.3.

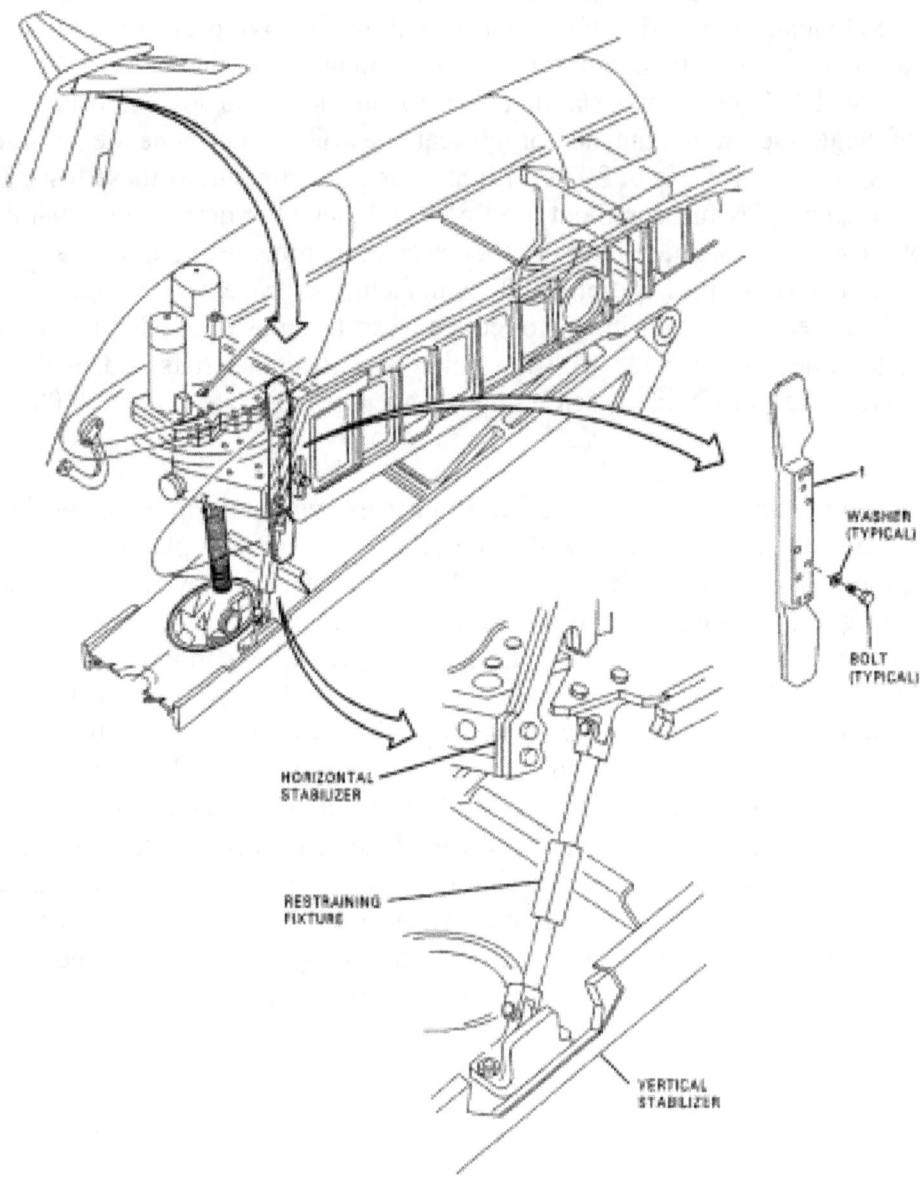

Figure 8. Depiction of how restraining fixture should be placed on the horizontal stabilizer during an end play check.

Figure 9. View of a typical dial indicator set up for end play checks.

On May 29, 1984, McDonnell Douglas issued AOL 9-1526, which reported that "two operators have reported three instances of premature removal/replacement of a horizontal stabilizer actuator assembly" and reiterated the OAMP document's recommendation that all DC-9 operators lubricate the jackscrew assembly at 600-flight-hour intervals. AOL 9-1526 added, "these assemblies, which had accumulated less than 6,000 flight hours each since new, were replaced due to excessive end play between the acme screw and acme nut." AOL 9-1526 stated that the jackscrew assemblies were returned to McDonnell Douglas for investigation and that "the acme nut installed in each assembly exhibited severe wear of the thread surfaces. In addition, grease samples taken from the lubrication passages…on two of these [jackscrew] assemblies were dry and without evidence of recent renewal. Accordingly, Douglas is of the opinion that the most probable cause of the observed acme nut thread wear and subsequent excessive end play was inadequate lubrication of the actuator assemblies."[84] The AOL concluded, "In view of the foregoing, Douglas wishes to emphasize the importance of maintaining a conscientious lubrication program to minimize acme nut thread wear and extend the service life of the actuator assembly."

In 1990, another airline reported that a DC-9 acme nut had worn beyond the 0.040-inch end play measurement limit after accumulating 5,780 flight hours, a wear rate of 0.0057 inch per 1,000 flight hours. In AOL 9-2120, dated December 6, 1990, McDonnell Douglas asked operators of DC-9 and MD-80 airplanes to report on the lubrication frequency of the jackscrew assembly, the mean time between removal (MTBR) and mean time between unscheduled removal (MTBUR), and the average wear rate of the assembly as measured by an on-wing end play check. On the basis of responses from 10 DC-9 operators and 11 MD-80 operators,[85] on September 5, 1991, McDonnell Douglas issued AOL 9-2120A, which provided the following results:

Model	Lube Interval	MTBR	MTBUR	Wear Rate
DC-9	1,329 hours	34,054 hours	34,395 hours	0.0011 inch per 1,000 hours
MD-80	804 hours	24,397 hours	28,397 hours	0.0013 inch per 1,000 hours

In AOL 9-2120A, McDonnell Douglas noted that the prematurely worn jackscrew assembly was not typical but, because the assembly was not examined by McDonnell Douglas, the cause of the premature wear was unknown.[86] However, McDonnell Douglas again reiterated its 600-flight-hour lubrication interval recommendations.[87]

On April 13 and November 20, 2000, Boeing issued revisions to the end play check procedure, including the following:

- Clarification of which part number (P/N) restraining fixture could be used on specific airplanes;

- Clarification of how and where to install dial indicator mounting bracket and probe;

---

[84] According to a McDonnell Douglas field service representative report dated March 13, 1984, the end play measurement of one of the jackscrew assemblies removed and returned to the company was 0.043 inch. The report stated that the acme nut threads were "worn and serrated." The report added that "it should be noted that lubrication cycle of jackscrew has been sporadic on this and other DC-9 [aircraft] at this location." The maintenance facility returned the assembly to McDonnell Douglas with a letter attached to the field representative report, which was addressed to product support and asked McDonnell Douglas to compare the returned assembly to drawing specifications and check the "trueness of the screw shaft assembly." The letter also asked, "What is the hardness spec of the nut?" The letter concluded, "I would appreciate an expeditious response in this matter for I have a later delivered aircraft with only 4,000 hours and within .001 of the maximum backlash [end play] limit."

[85] No information was available about reported wear rates for each of the operators that responded to the survey or for any individual jackscrew assembly that was included in the survey results. Therefore, the variability among operators or assemblies is unknown.

[86] Before issuing AOL 9-1526 in 1984, McDonnell Douglas examined six jackscrew assemblies that had been reported to have excessive wear. However, McDonnell Douglas could not examine the jackscrew assembly removed in 1990 because it was not delivered for inspection.

[87] The Safety Board confirmed that Alaska Airlines' maintenance department had received a copy of AOL 9-2120A from McDonnell Douglas.

- Addition of a note to specify that restraining fixture threads are to be cleaned and lubricated before the procedure is performed;

- Addition of a step to apply 100 inch-pounds to the restraining fixture in the lengthening direction and then 300 inch-pounds (instead of a range from 250 to 300 inch-pounds as previously stated in step [8]) on the restraining fixture in the shortening direction to get the end play; and

- Specification of the allowable end play limits in step (11) as "not less than 0.003 and not more than 0.040 inch" instead of "between 0.003 and 0.040."

1.6.3.3.2.1 Alaska Airlines' End Play Check Procedure

The Alaska Airlines maintenance task card for the acme screw and nut end play check, dated October 7, 1996, which was in effect at the time of the accident airplane's last end play check, described the end play check procedure as follows:

(1) Using control wheel trim switches, move horizontal stabilizer to approximately 1 degree airplane noseup position.

WARNING: TAG AND SAFETY OPEN CIRCUIT BREAKERS.

(2) Open the following circuit breakers.

Circuit Breaker	Panel Location	Panel Area
Autopilot & Alternate	upper epc	Left Radio
Longitudinal Trim (3)		AC Bus
Primary Longitudinal Trim (2)	upper epc	Left Radio DC Bus

(3) Remove panel and stabilizer fairing to gain access to work area.

(4) Install horizontal stabilizer restraining fixture.

(5) Clamp dial indicator mounting bracket to jackscrew torque tube restraining nut and position dial indicator probe against acme nut.

(6) Clamp dial indicator mounting bracket to upper stop on jackscrew and position dial indicator probe against lower plate of support assembly.

(7) Preload indicator probe to a least .100 inch and record dial indicator reading.

(8) Apply 250 to 300 inch pounds of torque to horizontal stabilizer by shortening restraining fixture and record dial indicator readings.

(9) Relieve torque on restraining fixture. Check that dial indicator has returned to initial reading.

(10) Repeat steps (8) and (9) several times to insure consistent results (within .001 inch).

(11) Check that end play limits are between .003 and .040 inch. Readings in excess of above are cause for replacement of the acme jackscrew and nut. Should replacement become necessary accomplish E.O. [engineering order] 8-55-10-01 if not previously accomplished.

(12) Check that free play between jackscrew and upper support does not exceed .010 inch. If free play exceeds this dimension, upper bearing must be replaced.

(13) Remove restraining fixture, dial indicator and attaching brackets and clamps. Install stabilizer fairing and panel.

(14) Close circuit breakers identified in step (2).

(15) Return horizontal stabilizer to neutral.

1.6.3.3.3 Horizontal Stabilizer Restraining Fixture

1.6.3.3.3.1 General

The McDonnell Douglas MD-80 generic task card for the acme screw and nut operation (No. 0855, dated December 1991) describes the end play check procedure and identifies a McDonnell Douglas-manufactured horizontal stabilizer restraining fixture, P/N 4916750-1, for use in the procedure. The task card also noted that equivalent substitutes could be used in place of these tools. The original Douglas engineering drawing for the restraining fixture indicates that the -503 configuration was to be used on MD-80 series airplanes, which would include the accident airplane.

1.6.3.3.3.2 Alaska Airlines-Manufactured Restraining Fixtures

The Safety Board's investigation revealed that up until the time of the accident, Alaska Airlines had only one restraining fixture in its inventory (at its OAK maintenance facility). The fixture was manufactured in-house; it was not manufactured by Boeing.[88] Since June 30, 1984, when Alaska Airlines began operating MD-80s, the restraining fixture was tracked through a general component tracking system as P/N 0-1301-0-0169 (manufacturer's P/N 4916750-1). There were no records of an initial inspection or any recurrent inspections of the restraining fixture. According to Alaska Airlines records, after the accident, it manufactured 11 additional restraining fixtures similar in design to its original fixture and purchased 7 fixtures manufactured by Boeing. The Alaska Airlines manager of tool control, who participated in the fabrication of the fixtures ordered by airline management after the accident, told Safety Board investigators in an interview that "what they [Alaska Airlines personnel] were making [the 11 restraining fixtures] wasn't even close" to Boeing's engineering drawing requirements.[89] He added that "we were directed to build the tools, and we did exactly what we were told."

During the investigation, Safety Board investigators conducted several on-wing end play checks using both restraining fixtures fabricated by Alaska Airlines[90] and those manufactured by Boeing. Examination of the restraining fixtures indicated that none of the fixtures manufactured by Alaska Airlines met Boeing's engineering drawing requirements for the fixtures. During several of these tests, the restraining fixtures fabricated by Alaska Airlines yielded lower end play measurements than the measurements obtained using the Boeing-manufactured fixture.[91] Figure 10 shows two of the restraining fixtures fabricated by Alaska Airlines and three of the fixtures manufactured by Boeing.

After the accident, on April 13, 2000, Boeing sent Message M-7200-00-00975 to all DC-9, MD-80, MD-90, and 717 operators to ensure that horizontal stabilizer inspection tooling conformed to the tool's engineering drawing requirements. Regarding the restraining fixture, the Boeing message stated the following:

> recent reports received by Boeing indicate that tooling not manufactured by Boeing is being utilized to perform the subject checks. In some cases, this tooling may not conform to the tool's drawing requirements...any variation in the tooling thread quality, pitch or amount of thread engagement can [a]ffect the wear check results.

The Boeing message asked operators to "ensure the restraining fixtures being utilized fully conform to the tool's drawing requirements" to "maintain consistent results from wear checks."

---

[88] In its submission, Alaska Airlines does not dispute that it had only one restraining fixture during this time. However, it contends that this fixture was manufactured by Douglas. As support for this contention, Alaska Airlines cites the July 11, 2002, deposition testimony (taken in connection with civil litigation arising from this accident) of the mechanic who conducted the September 1997 end play check of the accident jackscrew assembly that he thought the restraining fixture at the OAK maintenance facility in 1997 was "a Douglas part."

[89] Alaska Airlines did not have the Boeing engineering drawing requirements. The manager of tool control was given a copy of the engineering drawing for the in-house manufactured fixture for use in fabricating the additional 11 fixtures.

[90] The Safety Board did not use the actual restraining fixture that was used for the accident airplane's last end play check because it was not in serviceable condition.

[91] Discrepancies up to a maximum of about 0.005 inch were recorded.

| Factual Information | 46 | Aircraft Accident Report |

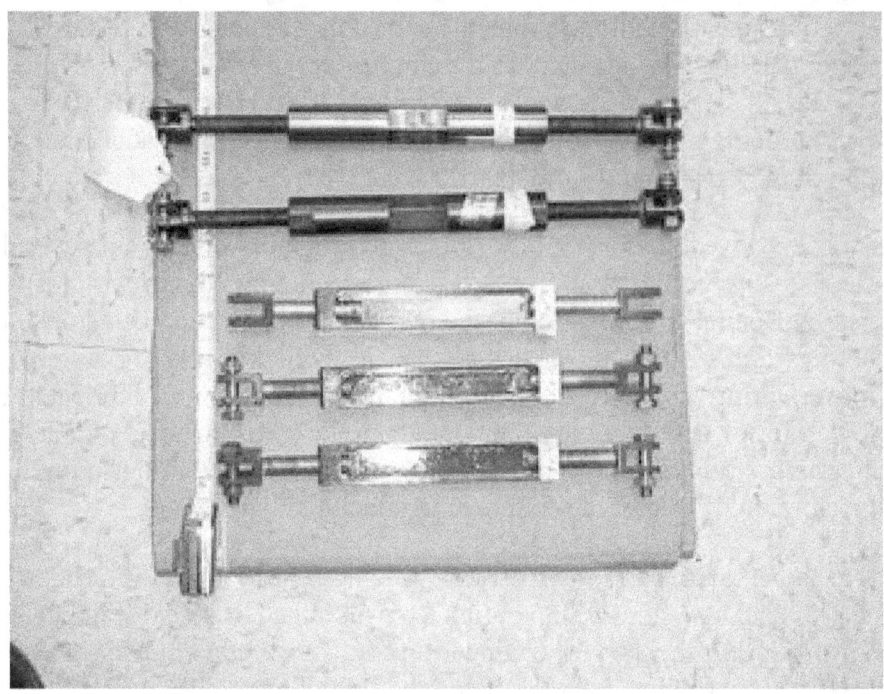

Figure 10. Two Alaska Airlines-fabricated restraining fixtures and three Boeing-manufactured fixtures.

On June 26, 2000, Safety Board investigators submitted a written request to Alaska Airlines for information about the restraining fixture it used during the September 1997 end play check on the accident airplane. The Board also requested information and documents regarding the physical properties of all of Alaska Airlines' in-house fabricated restraining fixtures. Investigators contacted Alaska Airlines on several occasions to follow up on these requests; however, Alaska Airlines did not provide the requested information until August 2, 2000.

On August 2, 2000, Alaska Airlines notified the FAA that it was concerned that the restraining fixtures it had manufactured in-house might not be "an equivalent substitute" for a Boeing fixture and, therefore, could produce erroneous measurements. Alaska Airlines thereafter suspended use of its in-house manufactured fixtures and temporarily grounded 18 of its 34 MD-80 series airplanes after determining that these airplanes might have received end play checks with the in-house manufactured tools. Before placing the airplanes back in service, Alaska Airlines performed end play checks on the grounded airplanes using Boeing-manufactured restraining fixtures; all of the rechecked assemblies had end play measurements within permissible limits. According to Alaska Airlines records, on August 4, 2000, it acquired 15 Boeing-manufactured restraining fixtures, all of which conformed to Boeing's engineering drawing requirements.

On December 4, 2000, the FAA issued Flight Standards Handbook Bulletin for Airworthiness 00-20A, "Equivalency of Special Equipment or Test Apparatus as Provided by Parts 43 and 145," which provides guidance for PMIs regarding the acceptability for equivalency of special equipment and/or test apparatus used in maintaining aircraft and their components. According to the FAA, the handbook bulletin was issued in connection with AD 2000-03-51, which referenced Boeing Alert Service Bulletin (ASB) DC9-27A362. ASB DC9-27A362, which specified the use of a restraining fixture for the required end play check, stated that the tool was currently unavailable from Boeing but that Boeing would supply the drawing necessary for operators to produce the tool. The ASB included eight items of guidance to PMIs for determining that repair stations and operators complied with the equivalency requirements, but it also stated, "[i]n instances involving the use of equivalent equipment and/or apparatus, the determination of equivalency for such equipment is the primary responsibility of the repair station or the air carrier, not the FAA."

1.6.3.3.4 End Play Check Intervals

1.6.3.3.4.1 Manufacturer-Recommended End Play Check Intervals

In the initial OAMP document for the DC-9, Douglas recommended that end play checks be performed at C-check intervals, or every 3,600 flight hours. Subsequently, in both the MSG-2 and MSG-3 MRB reports and OAMP documents for the DC-9 and MD-80 airplanes, the recommended end play check interval was increased to every other C check. Therefore, the MSG-2 OAMP document recommended that end play checks be performed every 7,000 flight hours, or 30 months, whichever comes first, and the MSG-3 OAMP document recommended that end play checks be performed every 7,200 flight hours, or 30 months, whichever comes first.[92]

1.6.3.3.4.2 Alaska Airlines' End Play Check Intervals

At the time of the flight 261 accident, Alaska Airlines' end play check interval was every 30 months, or about every 9,550 flight hours. The following is a chronology of acme screw and nut end play check intervals at Alaska Airlines:

- In March 1985, end play checks were scheduled for every other C check, which were accomplished every 2,500 flight hours; therefore, end play checks were to be accomplished every 5,000 flight hours.

- By July 1988, C-check intervals had been extended to 13 months, with no accompanying flight-hour limit; therefore, end play checks were to be accomplished every 26 calendar months, or about every 6,400 flight hours.

- In April 1996, C-check intervals were extended to 15 months, or about every 4,775 flight hours. Alaska Airlines' director of reliability and maintenance programs testified at the public hearing that this C-check interval extension was approved after the airline prepared a data package "that looked at five separate tail numbers [airplanes] that we selected as a stratified sample of the

---

[92] As stated previously, the MSG-2 recommended C-check interval was 3,500 flight hours, and the MSG-3 recommended C-check interval was 3,600 flight hours.

fleet. At that time I believe there were a little over 40 aircraft. We reviewed in detail all of the nonroutine discrepancies that had been written against the aircraft during those C checks." He indicated that individual maintenance tasks tied to C-check intervals (such as the end play check) were not separately considered in connection with the extension. However, he stated that based on the history of maintenance discrepancies noted for the five sample airplanes, two lubrication tasks were identified as inappropriate for extension along with the C check, and these tasks were converted to time-controlled (stand-alone) items to be performed at shorter intervals.[93] He indicated that all of the information was submitted to the FAA and that, approximately 1 month later, the extension was approved and implemented. The end play check interval remained at every other C check, which was then 30 calendar months, or about 9,550 flight hours.

- On April 6, 2000, after the accident, in accordance with ADs 2000-03-51 and 2000-15-15, the end play check became a time-controlled item required to be performed every 2,000 flight hours. Table 3 shows a comparison of manufacturer-recommended end play check intervals with Alaska Airlines' intervals.

Table 3. Comparison of manufacturer-recommended end play check intervals with Alaska Airlines' intervals.

Task Description	Original Douglas-Recommended Interval	MSG-2 MRB and OAMP	MSG-3 MRB	MSG-3 OAMP
C Check	3,600 flight hours	3,500 flight hours or 15 months, whichever comes first	3,600 flight hours or 15 months, whichever comes first	3,600 flight hours
End Play Check	C check (3,600 flight hours)	Every other C check (7,000 flight hours or 30 months, whichever comes first)	Every other C check (7,200 flight hours or 30 months, whichever comes first)	Every other C check (7,200 flight hours)

---

[93] The two tasks identified as requiring shorter intervals based on service history were (1) lubrication of the bent-up trailing-edge wing doors and (2) lubrication of bearings and bushings in the elevator hinges.

Task Description	Alaska Airlines 1985	Alaska Airlines 1988	Alaska Airlines 1996 to April 2000	Alaska Airlines Currently
C Check	2,500 flight hours	13 months (About 3,200 flight hours)	15 months (About 4,775 flight hours)	15 months
End Play Check	Every other C check (5,000 flight hours)	Every other C check 26 months (About 6,400 flight hours)	Every other C check 30 months (About 9,550 flight hours)	2,000 flight hours[a]

a. All carriers currently meet this requirement.

1.6.3.4 Accident Airplane's Maintenance Information

Alaska Airlines maintenance records documented that the following maintenance tasks and checks relevant to the jackscrew assembly were performed on the accident airplane:

- On May 27, 1993, a C check was accomplished at the OAK maintenance facility. The airplane had accumulated 2,674 flight hours. This C check included an end play check. Four nonroutine discrepancies were noted and corrected.

- On April 27, 1994, a C check was accomplished at the OAK maintenance facility. The airplane had accumulated 5,484 flight hours. No end play check was required, and no nonroutine discrepancies related to the horizontal stabilizer were noted.

- On May 17, 1995, a C check was accomplished at the OAK maintenance facility, including an end play check. The airplane had accumulated 9,194 flight hours. Five nonroutine discrepancies related to the horizontal stabilizer were noted. Maintenance action taken included cleaning the horizontal stabilizer jackscrew compartment around the vertical stabilizer and the areas around the right and left elevator hydraulic actuators.

- On June 20, 1996, a C check was accomplished at the OAK maintenance facility. The airplane had accumulated 12,906 flight hours. No end play check was required. Seven nonroutine discrepancies were noted and corrected. The records indicated that the elevators and horizontal stabilizer, including the jackscrew assembly, were lubricated.

- On February 27, 1997, according to maintenance records, the elevators and horizontal stabilizer, including the jackscrew assembly, were lubricated at the SFO maintenance facility.

- On September 26, 1997, the airplane entered a C check (which was completed on October 1, 1997) at the OAK maintenance facility, which included an end play check.[94] The airplane had accumulated 17,699 flight hours. The records indicated that the elevators and horizontal stabilizer, including the jackscrew assembly, were lubricated. The task card (No. 28312000) specified Mobilgrease 28 as the lubricant grease.

- On June 26, 1998, according to maintenance records, the elevators and horizontal stabilizer, including the jackscrew assembly, were lubricated at the SFO maintenance facility. The task card specified Aeroshell 33 as the lubricant grease.

- On January 13, 1999, a C check was accomplished at the OAK maintenance facility. The airplane had accumulated 22,407 flight hours. No end play check was required. No nonroutine discrepancies related to the horizontal stabilizer were noted. The records indicated that the elevators and horizontal stabilizer, including the jackscrew assembly, were lubricated. The task card specified Aeroshell 33 as the lubricant grease.

- On September 24, 1999, according to maintenance records, the elevators and horizontal stabilizer, including the jackscrew assembly, were lubricated at the SFO maintenance facility.[95] The task card specified Aeroshell 33 as the lubricant grease.

The Safety Board's review also revealed that, between July 4, 1999, and January 11, 2000, nine A checks were performed on the accident airplane. No nonroutine discrepancies related to the horizontal stabilizer were noted during any of these inspections. The accident airplane's last walk-around checks were performed at Anchorage International Airport, Anchorage, Alaska, and SEA on January 30, 2000. The airplane's last service check was also performed at SEA on January 30, 2000.

1.6.3.4.1 Accident Airplane's Last C Check Requiring an End Play Check

As stated previously, the accident airplane entered a C check, which included an acme screw and nut end play check (and several rechecks), at Alaska Airlines' OAK maintenance facility[96] on September 26, 1997. Completion of the C check and estimated time of release was scheduled for September 30, 1997, at 2300. According to Alaska

---

[94] For more details about the September 1997 C check, including the end play measurement, see section 1.6.3.5.1.

[95] Alaska Airlines maintenance records indicated that this was the last time that the accident airplane's jackscrew assembly was lubricated.

[96] The Alaska Airlines OAK maintenance facility operates 24 hours a day, 7 days a week. The work day is divided into the following three shifts: the graveyard shift, from about 2230 to 0630; the day shift, from about 0600 to 1430; and the swing shift, from about 1430 to 2300.

Factual Information 51 **Aircraft Accident Report**

Airlines records, on September 30, 1997, the estimated time of release was changed to October 1st at 2300. According to the October 2nd graveyard-shift turnover plan, the accident airplane was released on October 2nd at 2300. The following items were noted during the Safety Board's review of Alaska Airlines maintenance records for this C check:[97]

- September 27th (Saturday)—a nonroutine work card (MIG-4) was generated following an initial end play check on the accident airplane by a day-shift mechanic and an inspector. The MIG-4 noted the following discrepancy: "Horizontal Stab—acme screw and nut has maximum allowable end play limit (.040 in.)." The "planned action" box, which was filled out by the day-shift lead mechanic and inspector, stated, "Replace nut and perform E.O. 8-55-10-01." The swing-shift supervisor also signed off on the planned action. See figure 11.

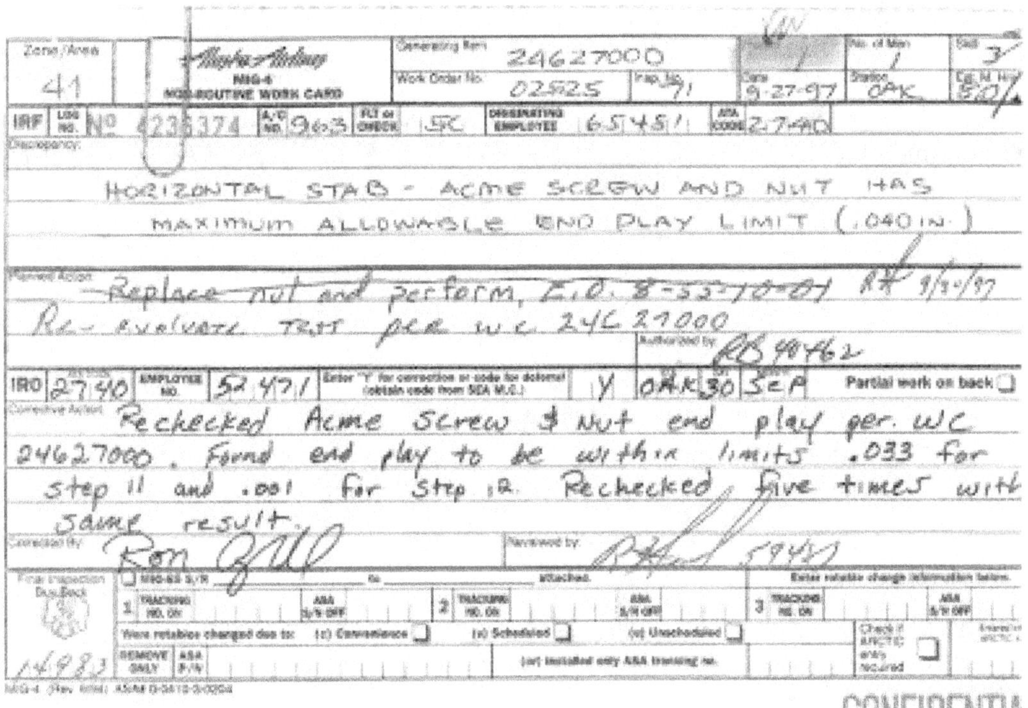

Figure 11. The September 27, 1997, MIG-4.

---

[97] All of the shift turnover plans were available except for the September 27 and 28, 2000, day-shift turnover plans. When asked by Safety Board investigators why these shift turnover plans were not available, Alaska Airlines responded that they were most likely not generated because of personnel problems at the time.

- September 28th (Sunday)—according to the day-shift turnover plan, the horizontal stabilizer was lubricated per task card No. 24312000 (which called for Mobilgrease 28).[98] The swing-shift lead mechanic made the following entry in the swing-shift turnover plan: "Ref MIG…horz stab:…get copy of E.O. & see what we need to do & order." The swing-shift turnover plan also contained the following entry related to the tail area: "N/R's [nonroutines], lube & treat C/W [complied with], still parts to order."

- September 29th (Monday)—the graveyard-shift lead mechanic made several entries in the graveyard-shift turnover plan, including "continue parts ordering (PANIC),"[99] "need copy of E.O. 8-55-10-01 for horiz stab acme screw and nut excessive end play," and "Tuesday [September 30th] departure looks doubtful." The day-shift turnover plan contained the following entry related to the tail area: "E.O. 8-55-10-01 needed for acme screw nut [excessive] end play." The swing-shift lead mechanic made the following entry on the swing-shift turnover plan: "re-do acme screw [check] to confirm problem…E.O. # is invalid.[100]…have [dayshift]…follow up."

- September 30th (Tuesday)—the graveyard-shift lead mechanic marked through the line on the MIG-4 that called for replacement of the jackscrew assembly and wrote, "re-evaluate test per WC [work card] 24627000." A different mechanic and inspector made an entry in the MIG-4's "corrective action" section, which stated, "Rechecked acme screw and nut end play per WC 24627000. Found end play to be within limits .033 for step 11 and .001 for step 12. Rechecked five times with same result." (See figure 11.) The MIG-4 was then reviewed by the graveyard-shift lead mechanic, who made the following entry in the graveyard-shift turnover plan: "stabilizer jackscrew acme nut C/W." The day-shift lead mechanic noted in the day-shift turnover plan, "tail close – I/W."[101] The day-shift turnover plan also noted, "partial O.K. to close up." The swing-shift turnover plan noted, "tail close (I/W)."

---

[98] Alaska Airlines maintenance records did not include the task card for the lubrication of the horizontal stabilizer. Because lubrication of the jackscrew assembly was a "repeated" maintenance item, the maintenance records only contained a complete task card for the most recent lubrication on September 24, 1999, at SFO.

[99] Alaska Airlines officials stated that jackscrew assemblies were not stocked as inventory items at the time of this C check and were only available through third-party vendors. Alaska Airlines' director of safety, quality control, and training told Safety Board investigators that "we do not believe a jackscrew was ordered. Our procedures call for the creation of a 'field requisition' whenever we order aircraft parts that are not in stock, such as the subject jackscrew." He added that field requisitions are noted in a field request log. He stated that a review of requisitions and logs related to the C check found no "jackscrew components listed in these documents. Based on this information, we do not believe a jackscrew or jackscrew components were ordered." He added, "we believe that the reference in the planned action to 'replace nut' caused some confusion since the nut should only be replaced as part of the jackscrew assembly…We also believe that the incorrect reference in the MIG-4 to perform E.O. 8-55-10-01 caused further confusion and interrupted the normal parts ordering process."

[100] The E.O. reference number was incorrect because it applied to another airplane model.

[101] "I/W" generally means "in accordance with."

- October 1st (Wednesday)—the graveyard-shift turnover plan noted, "[tail] close up needs work, 2 items closed without QA [quality assurance] OK."[102] The day-shift turnover plan noted, "tail closed up."
- October 2nd (Thursday)—the graveyard-shift turnover plan noted that the accident airplane "departed 0300 local time, so far so good."

1.6.3.4.2 Maintenance Personnel Statements Regarding the Accident Airplane's Last End Play Check

During postaccident interviews, the Alaska Airlines mechanic who conducted the end play check on September 27, 1997, stated that he did not recall what end play measurements were obtained, adding, "I don't remember the exact number, but it was below 40 [thousandths of an inch] I believe." He stated that he had previously accomplished about six end play checks and that "the ones I have done were well below the limits." He explained that during the procedure, "while I'm torqueing, he's [the inspector] looking at the dial indicator gauge." He also stated that he could not recall how many measurements were taken during the procedure. He stated that he noted the discrepancy on the MIG-4, adding that he did not consider the measurement to be beyond allowable limits. Referring to the decision to generate the MIG-4, he stated, "I don't really remember why. But, you know, maybe...maybe to have somebody take a look...you know, like a second opinion, maybe."

The September 27th day-shift inspector who was working with the mechanic (noting dial indicator readings) and who wrote the "planned action" instructions to replace the jackscrew assembly stated that he considered a 0.040-inch end play check result to be "at the end of the limit." He added, "It's not in excess. I would consider anything over 40 thousandths to be excess. So 40 thousandths would be right on the edge. How much further do you have to go before you're out of limits, especially if you got another two years before you're going to check it again?" The mechanic stated that he had never encountered similar wear during other end play checks and added the following:

> I've worked on every MD-80 that Alaska Airlines owns extensively. I have never come across any jackscrew that was worn to that extent. And we have aircraft, at that time we had aircraft that were 15 years old. We have what would be a relatively young aircraft [the accident airplane] that, for whatever reason, is, if it's at its limit or if it's at the end...it has a significant amount more wear than I've ever noticed in the past, and that's compared against aircraft that are three times its age.

The September 30th graveyard-shift lead mechanic who crossed out the planned action to replace the jackscrew assembly stated that he did so "because the MIG should not have been generated to begin with. There shouldn't have been a[n] MIG at all for the problem because it was within the allowable limits on the work card." He stated that he

---

[102] The Safety Board's investigation did not determine which two items were closed without quality assurance approval because Alaska Airlines records did not provide this information.

ordered a recheck of the end play "to take it one step beyond and prove that, in fact, it was as stated in the spec, 0.040 or less." He added the following:

> At that point in time, the planned action was inappropriate for the discrepancy. The discrepancy should have never existed. It fell within the work card limits, so, therefore, the discrepancy should have never been written. The planned action was inappropriate and…we could have opted to go ahead and sign it off right then and there. But I opted to re-evaluate to make sure that the limits stated were, in fact, either at 0.040 or below.

The inspector who worked with the mechanic who performed the recheck on September 30th stated that he had performed about three end play checks and never encountered results that were not within limits. He stated that the recheck "verified that the jackscrew was within limits." The mechanic who performed the recheck stated that all five end play measurements of 0.033 inch, as noted on the MIG-4, were "within a thousandths of it." He stated that the inspector and the lead mechanic were present during the recheck procedure.

1.7 Meteorological Information

According to the automatic terminal information service weather information recorded by the CVR at 1559:50, the following LAX weather conditions were current beginning about 1550:

> Wind two three zero at eight. Visibility eight. Few clouds at two thousand eight hundred. One two thousand scattered. Ceiling two zero thousand overcast temperature one six dew point one one. Altimeter three zero one seven. Simultaneous ILS [instrument landing system] approaches in progress runway two four right and two five left or vector for visual approach will be provided. Simultaneous visual approaches to all runways are in progress and parallel localizer approaches are in progress between Los Angeles International and Hawthorne airports. Simultaneous instrument departure in progress runway two four and two five.

1.8 Aids to Navigation

No problems with navigational aids were reported.

1.9 Communications

No external communication difficulties were reported.

1.10 Airport Information

Not applicable.

1.11 Flight Recorders

1.11.1 Cockpit Voice Recorder

The accident airplane was equipped with a Fairchild model A-100A CVR, S/N 62892. On February 2, 2000, the CVR was transported to the Safety Board's Washington, D.C., headquarters for readout and evaluation.[103] The exterior case was severely damaged by impact forces, but it had not been punctured. The tape and spool were wet but were otherwise intact and in good condition. The recording comprised four channels of good quality audio information:[104] one channel contained audio information recorded by the cockpit area microphone (CAM),[105] one channel contained audio information recorded by the public address system, and the other two channels contained audio information recorded through the radio/intercom selector panels at the captain and first officer positions. The CVR recording began about 1549:49 during cruise flight and ended about 1620:57, when the airplane impacted the water. A transcript was prepared of the 31-minute recording (see appendix B).

1.11.1.1 CVR Sound Spectrum Study

The Safety Board further examined the audio information recorded by the four CVR channels using a spectrum analyzer (which provides a visual representation of the frequency of the sound signals) and a computer signal analyzer (which allows analysis of the analog wave form and frequency content of the sounds and provides detailed timing information of the events).

The Safety Board analyzed the sound of two distinct "clicks" recorded on the CVR (referred to as "snaps" in the sound spectrum study report). The first click was recorded at 1549:54.8 and lasted about 0.01 second.[106] The first click was compared with a second

---

[103] The CVR was transported to Safety Board headquarters immersed in water in a sealed, plastic container.

[104] The Safety Board ranks the quality of CVR recordings in five categories: excellent, good, fair, poor and unusable. In a recording of good quality, most of the crew conversations can be accurately and easily understood. The transcript that was developed might indicate several words or phrases that were not intelligible. Any loss in the transcript can be attributed to minor technical difficulties or momentary dropouts in the recording system or to a large number of simultaneous cockpit/radio transmissions that obscure each other.

[105] The CAM is mounted in the overhead panel between the captain and first officer. The CAM is designed to capture sounds and conversations in the cockpit whenever the CVR is powered.

[106] According to the FDR, the autopilot disengaged about 1549:56.

click recorded at 1609:14.8 (just before the accident airplane's initial dive),[107] and the Board determined that the two sounds had identical frequency and energy characteristics.

The Safety Board also analyzed several distinct, very faint "thump" sounds recorded on the CAM channel. A set of thump sounds was recorded at 1609:16.9 (also just before the accident airplane's initial dive) and comprised two distinct thumps that were about 0.4 second apart. A series of distinct thumps was recorded at 1619:21.1 and comprised four thumps that were less than 0.1 second apart[108] and were recorded about 15 seconds before the second and final dive occurred at 1619:36.6. Analysis determined that all of the thumps had low-to-mid range frequency characteristics, which were not consistent with impulse sounds associated with metal-to-metal tapping or metal failure/tearing sounds. The frequency characteristics were determined to be similar to sounds produced by airflow.[109]

The Safety Board conducted a test to determine whether the failure of the acme screw torque tube would produce sounds consistent with those recorded on the CVR. In the laboratory, a hydraulic ram was used to pull the torque tube to failure, and the sounds associated with the torque tube failure were recorded.[110] The sounds recorded during the test were then compared with the sounds recorded on the accident airplane's CVR. Comparison of the data determined that the sound signatures recorded during the failure test did not match any of the sound signatures recorded by the accident airplane's CVR.

1.11.2 Flight Data Recorder

The accident airplane was equipped with a Sundstrand Data Control Universal FDR, S/N 9182. The FDR recorded information digitally on eight tracks using a 1/4-inch-wide magnetic tape that had a recording duration of 25 hours. On February 3, 2000, the FDR was transported to the Safety Board's Washington, D.C., headquarters for readout and evaluation. Although the FDR's outer protective case was severely damaged by high-impact forces, data were retrieved and analyzed.[111] The FDR began recording data from the accident flight about 1332 (after airplane startup and shortly before takeoff).

During its examination of the data, the Safety Board determined that two parameters, angle-of-attack and left brake-pressure, were not functioning properly and were recording faulty engineering values.

---

[107] According to the FDR, the autopilot disengaged about 1609:16.

[108] These four thump sounds were referred to as the "sound of faint thump" in the CVR transcript.

[109] Although the Safety Board could not determine the direct source of the sounds, they were similar to sounds recorded during previously conducted flight tests involving wake vortex encounters.

[110] The data were collected using a calibrated sound pressure transducer and an accelerometer attached to a mount used to hold the torque tube in place during the test.

[111] For information on airplane performance studies based on an analysis of the FDR data, see sections 1.16.1 and 1.16.1.1.

1.12 Wreckage and Impact Information

1.12.1 General

The on-site phase of the investigation, including wreckage recovery, examination, and documentation, was accomplished between February 1 and March 15, 2000.[112] Throughout most of the on-site investigation, remote-operated vehicles and side-scan sonar equipment were used to search the underwater debris field.[113] The majority of the debris field was cleared using the remote-operated vehicles to either load baskets with, or wrap cables around, the wreckage. In the late stages of wreckage recovery, a commercial trawler was used to drag the ocean bottom and recover smaller wreckage pieces.[114] All of the recovered wreckage was unloaded at Port Hueneme, California, for examination and documentation.

About 85 percent of the fuselage was recovered. The largest sections were from below the airplane's floor line. The fuselage's upper crown structure was broken into smaller pieces and had substantial compression damage. The recovered fuselage segments increased in size from the forward section of the airplane (where they were the smallest) to the aft section of the airplane (where they were the largest).

No evidence indicated any pre- or postimpact fire damage. Fractures found on pieces of fuselage wreckage were consistent with failures generated by high-energy impact. Examination of the fracture surfaces found no evidence of preexisting cracking or foreign impact damage.

The majority of both wings were recovered. The right wing exhibited more damage than the left wing.

About 85 percent of the empennage, which comprises the horizontal and vertical stabilizers and the engines, was recovered. Most of the horizontal stabilizer was recovered, including the center box and the left and right outer sections. All of the empennage's control surfaces were recovered except for a 4-foot section of the left elevator and the geared tab.

The Safety Board examined the horizontal stabilizer wreckage for evidence of preimpact failures and structural damage that might have occurred during the accident flight. Damage and/or witness marks on the horizontal stabilizer were visually examined to determine what degree of horizontal stabilizer movement (travel) would have been necessary to create the witness marks or damage.

---

[112] Floating wreckage debris was recovered throughout the on-site phase of the investigation.

[113] Side-scan sonar searching preceded the underwater wreckage recovery to determine the boundaries of the underwater debris field, which was spread over a 1/4-mile area of the ocean bottom at a depth of about 700 feet.

[114] About 15 percent of the wreckage was recovered using this method. Ten percent of the wreckage was not recovered.

The horizontal stabilizer was found fragmented into several pieces. The right side of the horizontal stabilizer was more fragmented than the left side. A 1.5-foot section of the vertical stabilizer was found attached to the horizontal stabilizer's pivot fitting lugs. The bottom side of both lugs were scraped, and the grease fittings located on the bottom side of each lug were found sheared off. Examination determined that the leading edge of the horizontal stabilizer would have to rotate up to a position about 16.2° leading edge up for the grease fittings to sheer off. The adjacent vertical stabilizer pivot structure had contact marks (in line with the grease fitting location) consistent with the leading edge of the horizontal stabilizer having rotated to a position of about 80°. The vertical stabilizer was found fragmented, consistent with high-energy impact damage. The vertical stabilizer's tip fairing was recovered broken into two segments. The recovered segments of the vertical stabilizer's tip fairing were placed together, and the two fracture surfaces were matched. The forward 25 percent of the vertical tip fairing was not recovered. Examination of the matched fairings portions showed upward deformation damage that was also consistent with the leading edge of the horizontal stabilizer having rotated to a position of about 80°.

The horizontal stabilizer front spar lower cap had a 1.5-inch semicircular indentation about 3.5 inches left of the centerline, consistent with contact by the acme screw. The jackscrew assembly side attachment plates were bent about 1.25 inches to the left, and the side plate lugs were partially fractured, consistent with high-energy impact damage.

Both engines were recovered separated. The left (No. 1) engine was found separated into three sections at the high-speed compressor.[115] No evidence indicated an engine fire or an uncontained engine failure. Examination of the recovered compressor groups from the left engine revealed marks and signatures consistent with the engine producing significant rotational energy at impact. The thrust reverser was found in the stowed position.

The right (No. 2) engine was recovered in one piece except for the entire fan front and rear cases, which were not recovered. None of the recovered fan blades exhibited significant blade tip rubbing. No evidence indicated integral fan hub and disk distress or rupture. No evidence indicated an engine fire or an uncontained engine failure. The recovered compressor blades and vanes did not exhibit signs of distress from engine rotation. Examination of the recovered compressor groups from the right engine revealed signatures consistent with the engine producing little rotational energy at impact. The thrust reverser was found in the stowed position.

1.12.2 Horizontal Stabilizer Jackscrew Assembly Components

The main supporting elements of the horizontal stabilizer jackscrew assembly were found intact and attached to the horizontal stabilizer's front spar. The acme screw was found cracked but attached to the support assembly. Metallic filaments were found

---

[115] The high-speed compressor was not recovered.

Factual Information 59 **Aircraft Accident Report**

wrapped around the central part of the acme screw.[116] (Figures 12a and 12b show the acme screw immediately after it was brought on board the recovery ship and immediately after it was brought to shore, respectively; figure 12c shows the acme screw during initial inspection.) The acme screw was rinsed with fresh water immediately after its recovery by directing the water from a hose similar to a garden hose over the gearbox and allowing the water to flow down over the screw. To preserve the acme screw and the thread remnants wrapped around the screw in their as-recovered condition, the stream of water used to rinse the screw was not directly sprayed on it. The acme screw might have been rinsed again with a garden-type hose after it was brought to shore.

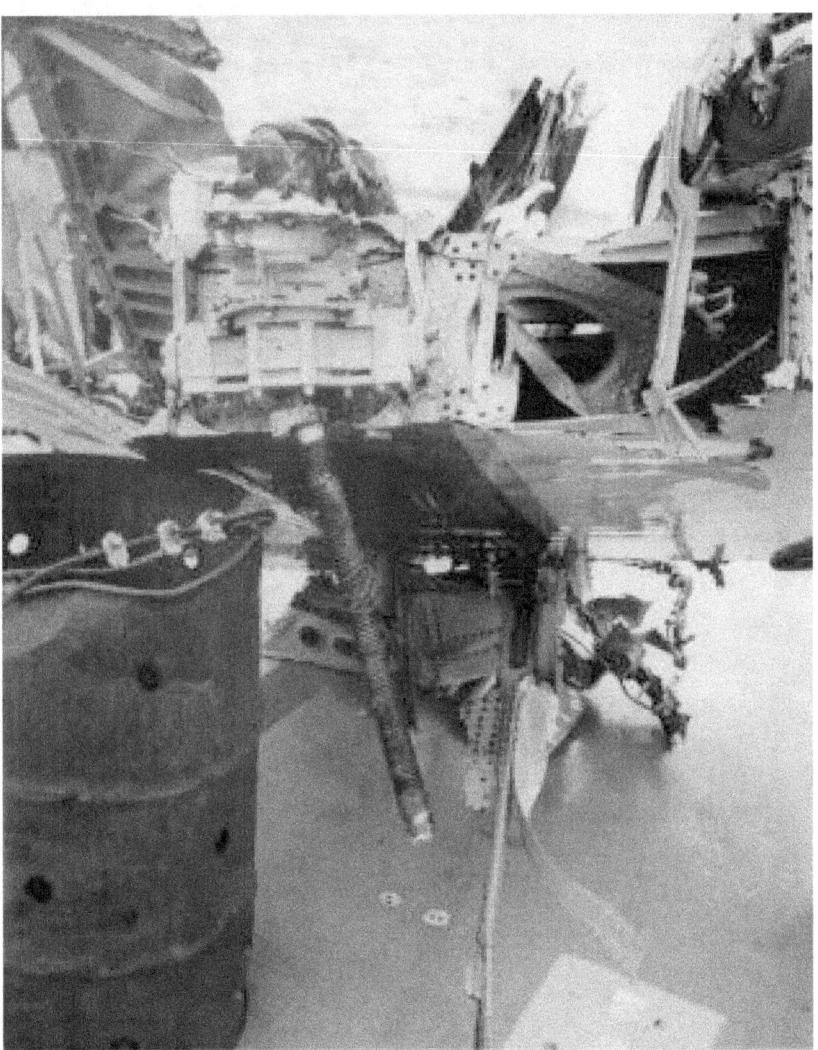

Figure 12a. The acme screw immediately after it was brought on board the recovery ship.

---

[116]These metallic filaments were later identified as acme nut thread remnants. After initial mechanical and metallurgical examinations, the acme screw, thread remnants, and acme nut were removed from the surrounding structure and sent to the Safety Board's Materials Laboratory for examination. For details about the metallurgical examinations of these components, see section 1.16.2.

Figure 12b. The acme screw immediately after it was brought to shore.

Figure 12c. The acme screw during initial inspection.

The upper mechanical stop was found attached to the splines of the acme screw, and the clamp bolt for this stop was in place. The lower mechanical stop and its clamp bolt were recovered as a separate unit from the acme screw. Several impact/scrape marks were found on the upper surface of the lower mechanical stop, and a round impact mark was found on the lower surface of the stop. Evidence of damage to the splines on the inside diameter of the lower mechanical stop was observed.[117]

The portion of the torque tube that extended beyond the lower end of the acme screw was fractured. The lower end of the torque tube, the torque tube nut washer, and the torque tube nut were not recovered.

The acme nut and gimbal assembly were found attached in the vertical stabilizer structure. The gimbal assembly was found to be movable on both rotational axes, except in areas distorted by surrounding structure.

The primary trim motor was found detached from the differential gearbox, but it was mostly intact. The alternate gearbox was found attached to the primary motor along with a portion of the upper housing of the differential gearbox. Both electrical connectors to the primary motor were present; however, the connection wires were missing.

---

[117] For more information about the damage to the splines on the inside diameter of the lower mechanical stop, see section 1.16.2.4.

All recovered pieces of the accident airplane's primary trim motor were examined. No evidence of any thermal damage was noted. The primary trim motor's auxiliary gearbox was removed from the motor, disassembled, and examined. No evidence indicated any preimpact mechanical damage.[118] All recovered pieces of the accident airplane's alternate trim motor were examined. Impact damage and gouging were predominantly located toward the left rear portion of the motor. No evidence indicated any preimpact mechanical damage.

The primary up-trim relay, the down-trim relay, the up-trim brake relay, the down-trim brake relay, and the trim brake override switch were examined. No evidence of thermal damage or excessive wear was observed on either the down-trim or the up-trim relay contacts. Corrosion was observed on all metal components of both trim relays. No evidence of any preimpact mechanical malfunction was found on either relay. Examination and testing of the up-trim and down-trim brake relays found no evidence of any preimpact mechanical malfunction or abnormal wear. An electrical continuity check of the trim brake override switch determined that it was in the neutral position. No evidence indicated any preimpact mechanical malfunction.

1.13 Medical and Pathological Information

Tissue specimens from both pilots were transported to the FAA's Civil Aerospace Medical Institute laboratory for toxicological analysis. The laboratory performed its routine analysis for major drugs of abuse[119] and prescription and over-the-counter medications, and the results were negative. Although the analysis detected ethanol in the tissue specimens of both pilots, the analysis report noted that the ethanol found was consistent with postmortem ethanol production.

1.14 Fire

There was no evidence of an in-flight fire.

1.15 Survival Aspects

The airplane was destroyed by high-impact forces, and no occupiable space remained intact. The Ventura County, California, Coroner's report stated that all of the accident airplane's occupants died as a result of blunt force impact trauma.

---

[118] According to the manufacturer of the primary trim motor (Sunstrand), it had overhauled the accident airplane's primary trim motor on February 22, 1994, and again on October 4, 1996.

[119] The five drugs of abuse tested are marijuana, cocaine, opiates, phencyclidine, and amphetamines.

1.16 Tests and Research

1.16.1 Airplane Performance

The Safety Board used FDR, CVR, ATC transcript, and radar[120] data to determine the accident airplane's motions and performance during the accident sequence. These data (and associated calculations) indicated the following:

- The last horizontal stabilizer movement before the initial dive was recorded at 1349:51 (about 2 hours and 20 minutes before the initial dive) during the climbout from PVR when the airplane was climbing through approximately 23,400 feet at 331 KIAS. At that time, the horizontal stabilizer moved from 0.25° to 0.4° airplane nose down.

- Following the horizontal stabilizer movement at 1359:51, the airplane continued to climb at 330 KIAS. The airplane began to level off at approximately 26,000 feet and slowed to 320 KIAS. At 1351:51, the airplane began to climb again, and, by 1352:50, it had slowed to about 285 KIAS as it climbed to approximately 28,600 feet and then leveled off.

- At 1353:12, the airplane was at approximately 28,557 feet and 296 KIAS, and the autopilot disengaged.

- The airplane began to climb at a much slower rate over the next 7 minutes. During this part of the ascent, the elevators were deflected between -1° and -3°, and the airplane was flown manually using up to 50 pounds of pulling force.

- By 1400:00, the airplane had reached an altitude of approximately 31,050 feet. For about the next 24 minutes, the airplane was flown manually using approximately 30 pounds of pulling force while maintaining level flight at 280 KIAS. Starting at 1424:30, the airspeed increased, eventually reaching 301 KIAS,[121] at which time the airplane was flown using about 10 pounds of pulling force. At 1546:59, the autopilot was re-engaged.

- At 1609:16, the autopilot disengaged, and the CVR recorded the sound of a clunk. At 1609:16.9, the CVR recorded the sound of two faint thumps in succession. During the 3 to 4 seconds after the autopilot was disengaged, the horizontal stabilizer moved from 0.4° airplane nose down to a recorded position of 2.5° airplane nose down,[122] and the airplane began to pitch down, starting a dive that lasted about 80 seconds as the airplane went from approximately 31,050 feet to between 23,000 and 24,000 feet. During the

---

[120] Air Route Sur\veillance Radar data were obtained from the FAA's Los Angeles ARTCC, and Airport Surveillance Radar data were collected from several airport facilities in the Southern California Terminal Radar Control area and from the U.S. Air Force 84th Radar Evaluation Squadron.

[121] The flight plan called for a cruise speed of 283 knots calibrated airspeed.

[122] The horizontal stabilizer remained at the recorded position of 2.5° until the last minute of the flight. For details about simulation studies conducted by the Safety Board to determine the actual position of the horizontal stabilizer during this time, see section 1.16.1.1.

movement of the horizontal stabilizer, the CVR recorded the sound of the stabilizer-in-motion tone for 3 to 4 seconds. During the next several seconds, the load factor oscillated between 0.4 and 1.0 G, and the elevators moved to the trailing-edge-up direction from -0.4° to -7.0°,[123] consistent with control inputs being made to bring the nose up.

- By 1609:42, the airplane had descended to approximately 29,450 feet and had accelerated to about 322 KIAS. At this time, the airplane's pitch angle was -7.9° airplane nose down, and the elevators had moved to about a -8° position. The speed brakes deployed 1 second later and remained deployed for the next 1 minute 30 seconds.

- At 1610:00, the airplane was accelerating through 343 KIAS (the airplane's maximum allowable airspeed). At 1610:01.9, the CVR recorded the sound of the overspeed warning.

- By 1610:16, the airplane had accelerated to 353 KIAS, the maximum airspeed it reached during the initial dive, and this airspeed remained nearly constant until 1610:23 as the airplane was descending through 24,200 feet. By this time, the airplane's pitch angle had increased from -8° to -5° airplane nose down, and the elevators had moved to about a -11° position.

- About 1611:13, the speed brakes were stowed. By this time, the airplane's airspeed had decreased to 262 KIAS, and the airplane was maintaining an altitude of approximately 24,400 feet with a pitch angle of 4.4°. During the next 10 seconds, the airplane's airspeed decreased to 254 KIAS. The airplane remained between 23,000 and 24,000 feet for the next 5 minutes, and the airspeed gradually increased to 335 KIAS. During this period, elevator movements ranged from between -7° to -9°. Airplane performance calculations indicated that a control column force of between about 130 and 140 pounds was required to recover from the dive.[124]

- At 1614:44, the speed brakes deployed again, and the airplane's airspeed began to decrease.

- At 1617:50, the speed brakes were being stowed. By this time, the airplane had leveled off at approximately 18,000 feet, and the airspeed had decreased to 253 KIAS. At this time, the elevators were operating between -10° and -13°, and the airplane's pitch angle was between -2° and 0°, with a constant indicated horizontal stabilizer angle of 2.5° airplane nose down.

- At 1617:54, the CVR recorded the captain order deployment of the slats. About 1618:00, when the airplane was passing through 17,800 feet at 252 KIAS, the FDR recorded the extension of the leading edge slats. About 1618:05, the CVR recorded the captain order deployment of the flaps. About 3 seconds later, the flaps began extending to 11°, and the airspeed

---

[123] The available range of the airplane's elevators is from +15° to -25°.

[124] As stated previously, this estimated force could be applied by either the captain, the first officer, or both.

decreased about 4 knots. The flaps remained at this value for about 20 seconds, during which time the airplane's pitch angle was about 0°, with changes of less than 1°. About 1618:26, the CVR recorded the captain order retraction of the slats and flaps. During the 50 seconds after retraction of the slats and flaps, the airplane's indicated airspeed increased from 245 to 270 knots, and the airplane climbed from approximately 17,400 to 17,900 feet with a pitch angle of about 4°.

- Beginning at 1619:09, the FDR recorded small oscillations in elevator angle, pitch angle, and vertical acceleration values for about 15 seconds.

- At 1619:21, the CVR recorded the sound of a faint thump. At 1619:21.5, the FDR recorded a 0.09° change, from 2.51° to 2.60°, in the horizontal stabilizer position (this was the first change in recorded horizontal stabilizer position since the one that coincided with the airplane's initial dive from 31,050 feet about 10 minutes earlier) and elevator deflections increasing from -9° to -13°. Pitch angle and elevator oscillations ceased 2 seconds after the stabilizer position changed.

- At 1619:29, the CVR recorded the captain order redeployment of the slats and flaps. At 1619:35, when the airplane was at an altitude of 17,900 feet, an indicated airspeed of 270 knots, and a pitch angle of 2°, the flaps were transitioning from 7° to 11°.

- At 1619:36, the left elevator value was -12°, the right elevator value was -11°, and the load factor was 1 G. At 1619:36.6, the CVR recorded an extremely loud noise, and the FDR recorded a right elevator position of -25°.[125] By 1619:36.6, the load factor had increased to 1.45 Gs. At 1619:36.9, the pitch angle began to change to the airplane-nose-down direction, and the load factor decreased to 0.67 G. At this time, radar data detected several small primary returns. Within 3 seconds, the load factor quickly decreased to -3 Gs. The next several seconds of FDR data indicated a maximum airplane-nose-down pitch rate of nearly 25° per second.

- By 1619:40, the airplane was rolling left wing down, and the rudder was deflected 3° to the right.

- By 1619:42, the airplane had reached its maximum valid recorded airplane-nose-down pitch angle of -70°, and the load factor had increased to -1.45 Gs. At this time, the roll angle was passing through -76° left wing down. During the next 2 seconds, the load factor decreased to -2.1 Gs.

- By 1619:45, the load factor had increased to -1.75 Gs, the pitch angle had increased to -28°, and the airplane had rolled to a -180° inverted position. Further, the airplane had descended to approximately 16,420 feet, and the indicated airspeed had decreased to 208 knots.

---

[125] All subsequent recorded elevator values exceeded the physical ranges of the airplane's elevator system and were considered invalid.

- At 1619:54, the left aileron moved to more than 16° (to command right wing down), and, during the next 6 seconds, moved in the opposite direction to -13° (to command left wing down).

- At 1619:57, the rudder was returned to the near 0° position, the flaps were retracted, and the airplane was rolling through -150° with an airplane-nose-down pitch angle of -9°.

- After 1619:57, the airplane remained near inverted, and its pitch oscillated in the nose-down position until it impacted the ocean. FDR data indicated large rudder and aileron movements during the final minute before impact. Further, engine pressure ratio[126] parameters fluctuated erratically during the final dive, which is consistent with airflow disturbances associated with extreme angles-of-attack.

1.16.1.1 Airplane Performance Simulation Studies

The Safety Board conducted several performance studies to reproduce, in simulation models, the flight data parameters on the accident airplane's FDR using the same control inputs recorded on the FDR (for example, aileron, elevator, horizontal stabilizer angle, and engine torque). Additionally, kinematics studies[127] using MD-80 aerodynamic data obtained from Boeing calculated the control deflections required to match the airplane response on the FDR. These calculations provided an estimate of control inputs for use in simulations when control inputs recorded on the FDR did not produce the same response in the simulations as those recorded on the FDR. These kinematics calculations also provided the tension load on the jackscrew assembly.

The airplane performance studies focused on the airplane's initial dive (about 1609:16), which occurred after the autopilot was disengaged, and the second and final dive (about 1619:37), which occurred after an extremely loud noise was recorded by the CVR. Results of simulation and kinematics studies indicated that, before the autopilot was disengaged, the values recorded for the horizontal stabilizer, elevator, aileron, and rudder produced the same airplane responses as those recorded on the FDR during cruise flight. After the autopilot was disengaged and the airplane subsequently pitched down, the simulation results showed considerable discrepancies between FDR data and simulated airplane response. With the use of the recorded horizontal stabilizer angle, simulations were conducted using control inputs based on those recorded by the FDR. The simulations showed that the airplane would have recovered quickly and climbed using the stabilizer angle and elevator positions and movements recorded on the FDR, indicating that recorded values of the longitudinal controls were not accurate representations of the actual positions on the accident airplane after the initial dive.

---

[126] Engine pressure is a measure of engine thrust, comparing total turbine discharge pressure to the total pressure of the air entering the compressor.

[127] Kinematics is a process that involves fitting curves through available FDR data (such as heading, pitch, and roll) and calculating accelerations, forces, and moments from these rates. The flight control surface positions required to match the FDR angular and position data are then calculated using aerodynamic data for the airplane that have been provided by the manufacturer.

Kinematics studies were conducted to account for the discrepancies between the FDR data and the simulation results. One study assumed that the stabilizer value was accurate and calculated the elevator movement required to match the FDR-recorded values. A second study assumed that the elevator values were accurate and calculated the stabilizer movement required to match the FDR-recorded values. Simulations indicated favorable matching with FDR values using a constant, increased offset in the horizontal stabilizer position compared with the recorded elevator position. On the basis of these findings, further kinematics and simulator studies assumed that the elevator positions recorded by the FDR after the initial dive were accurate and that the recorded horizontal stabilizer positions recorded by the FDR were inaccurate.

Results of kinematics and simulation studies indicated that the accident airplane's performance was consistent with the horizontal stabilizer position moving beyond the 2.5° airplane-nose-down value recorded on the FDR during the initial dive (causing the airplane to pitch nose down as recorded, assuming the elevator positions were as recorded on the FDR). Kinematics study results indicated a relatively steady and consistent airplane-nose-down horizontal stabilizer angle in the 10 minutes between the initial dive and the second dive, followed by a large increase in stabilizer angle at about the time valid elevator data were lost on the FDR and the loud noise was recorded on the CVR. Calculated tension loads on the acme screw between the first and second dives were considerably larger than the loads encountered in the normal operational envelope. The studies indicated that the performance of the accident airplane during the second dive was consistent with the horizontal stabilizer moving quickly to a higher leading-edge-up position (which corresponds to airplane nose down) of at least 14° just after the loud noise was recorded by the CVR.

1.16.2 Metallurgical Examinations

The Safety Board's metallurgical investigation included on-scene and laboratory examinations. Initially, recovered wreckage was examined at a dockside facility at Port Hueneme, California, from February 9 to 11, 2000. Metallurgical examinations focused on the horizontal stabilizer, the jackscrew assembly, and related components. After the on-scene examinations concluded, laboratory examinations were conducted at the Safety Board's Materials Laboratory, Washington, D.C.; the Boeing Commercial Aircraft Materials and Process Engineering Laboratory, Long Beach, California; and Integrated Aerospace, Santa Ana, California.

1.16.2.1 Horizontal Stabilizer Forward Spar

The jackscrew assembly (not including the lower mechanical stop, the lower end of the torque tube, the torque tube retaining nut and washer, and the trim motors) was recovered attached to the front of the horizontal stabilizer forward spar.[128] Spar brackets through which the acme screw attachment bolts pass were found severely bent. The lower

---

[128] The jackscrew assembly is attached to spar brackets on the front of the horizontal stabilizer by six bolts, three on each side of the spar.

legs of both brackets were fractured at bend locations. A visual inspection determined that the fractures were consistent with overstress separations.

1.16.2.2 Acme Screw

Several contact marks were found on the acme screw surface. A downward-facing, C-shaped contact mark on cadmium plating was found overlaying one end of a crack at the upper end of the acme screw. An inspection of the adjacent horizontal stabilizer structure found a mating dent in the horizontal stabilizer forward spar's lower flange. The spar flange dent was smoothly curved with a radius that matched the acme screw at the C-shaped contact mark location. A visual inspection also determined that the upper end of the acme screw had contacted the bore of the pass-through hole in the lower gearbox support plate. Marks indicated several major impacts at various aft positions and minor impacts at the forward positions within the bore.

Corrosion pitting and red rust areas were found along most of the acme screw's length, with the heaviest areas of corrosion and rust along its lower half. Large areas of the acme screw were also found covered by white deposits, which chemical analysis determined were consistent with corrosion debris from the magnesium gearbox case attached to the top of the acme screw assembly. Figure 13 shows the acme screw thread remnants wrapped around the screw, the red rust areas, and the white deposits.

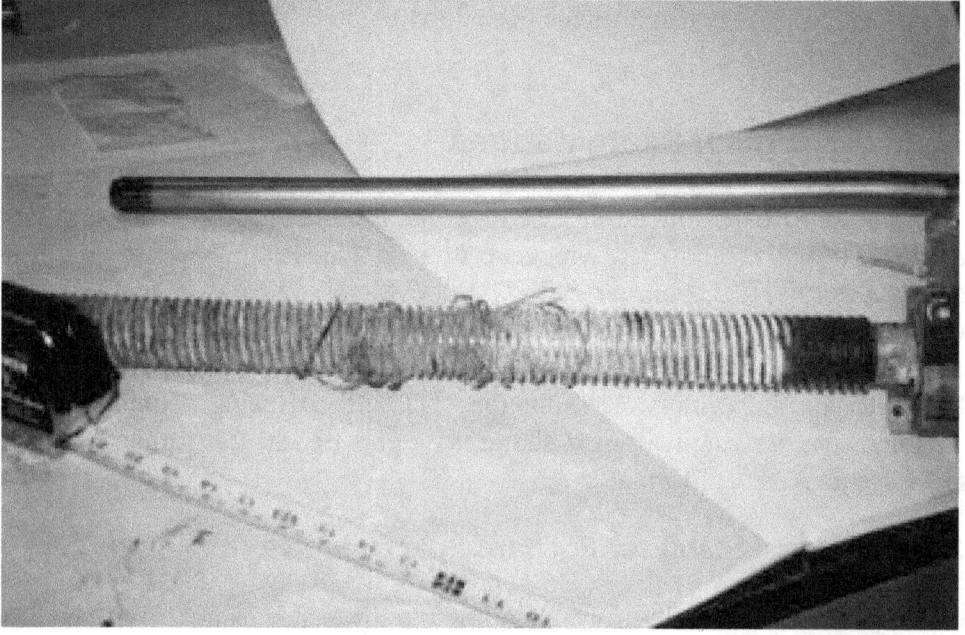

Figure 13. A photograph of the acme screw thread remnants wrapped around the screw, the red rust areas, and the white deposits.

On-scene visual and tactile inspections by Safety Board metallurgists of the acme screw's threaded areas found no evidence of grease in any condition, either semi-fluid (that is, fresh) or solid/dry (that is, old or degraded), or other lubricants in the central "working region" of the screw threads.[129] Laboratory examinations found small flakes of dried and hardened grease attached to some of the thread remnants in this region. The acme screw's lower threads (which are outside of the working region) were found partially packed with a mixture consistent with sand and grease. (Figure 14 shows the sand/grease mixture packed between the acme screw's lower threads.) Parts of the acme screw's upper six to eight threads had an oily sheen, and small deposits of greaselike material were found between the threads.

Figure 14. A photograph of the sand/grease mixture packed between the acme screw's lower threads.

---

[129] The acme screw working region is that part of the screw that can come in contact with the acme nut during its operation between the upper and lower electrical stop limits.

The acme screw was locally bent about 5° at the point of a crack in the area of the C-shaped contact mark above the acme screw threads and below the upper stop splines. The crack was gapped open about 1/16 inch and extended about 270° around the acme screw's circumference. An initial visual inspection of the crack surfaces and subsequent microscopic laboratory examination determined that the crack was on a 45° plane, consistent with an overstress fracture. No evidence indicated any preexisting cracks.

The acme screw was cleaned with soap and water for more detailed laboratory examination. Cleaning removed most of the dried and hardened grease and some of the white magnesium deposits; however, most of the red-rusted, corrosion-pitted areas remained. Laboratory examination revealed that there was less pitting at the upper and lower parts of the acme screw where the screw's black-oxide coating was more intact.[130] No wear was visible on the flanks of the acme screw's 10 lower and 7 upper threads. The upper flanks of the threads in the working region showed evidence of polishing and removal of the black-oxide coating that increased toward the center of the acme screw.

Examination of the surfaces through the black-oxide coating found no machining steps (indentations worn into the surface) or sharp features. Thread flank surfaces in areas that did not have a black-oxide coating appeared circumferentially worn and lustrous with no wear steps or sharp features found in these areas. Surface roughness was measured at four locations on the flanks of the acme screw threads, and all locations recorded a surface finish of 32 root mean squared (RMS) value[131] or finer, as required by the manufacturer's specifications.

The lower end of the acme screw, including about 19 threads, was cleaned electrolytically,[132] which completely removed the corrosion debris and most of the white magnesium deposits but left the black-oxide coating, paint, and cadmium plating (on the splined areas) intact. Subsequent microscopic examination determined that corrosion had deeply pitted the acme screw surface. Surface finish was measured on the thread flank in the cleaned lower end and determined to be 32 RMS or finer. Chemical and metallurgical tests established that the acme screw met material and process specifications.

1.16.2.3 Acme Screw Torque Tube

The acme screw torque tube was fractured approximately in line with the bottom of the screw, through the first full thread just below the external splines at the lower end of the tube. Neither the mating lower piece of the torque tube nor the retaining nut and washer from this area of the assembly were recovered. The upper end of the torque tube

---

[130] A black-oxide coating is a thin surface finish treatment that provides limited corrosion and wear protection.

[131] RMS is a numerical measurement of surface roughness. The acme screw's engineering drawing specifications require a surface finish of 32 RMS or finer in the threaded area. In later specifications, RMS has been replaced by R_A (arithmetic average).

[132] Electrolytic cleaning is accomplished by applying a small amount of voltage to the part while it is immersed in a mildly caustic solution.

was bent about 5° in a location corresponding to the location of the crack in the acme screw.

Macroscopic fracture traces indicated that the torque tube fracture initiated from a broad front at the tube's right forward quadrant (with the torque tube oriented as recovered) and propagated through the tube's cross-section. Macroscopic fracture features emanated from a thin reflective fracture band adjacent to the thread root that was determined to be a preexisting fatigue crack. Fatigue striations were identified throughout the band, consistent with low-cycle fatigue cracking.[133] The fatigue band followed the first thread root around almost 330° of the torque tube's circumference and extended about 0.03 inch onto the fracture surface in the initiation area. The remaining fracture area was ductile overstress stemming from the fatigue. Chemical and metallurgical tests established that the torque tube material met material and process specifications.

The large diameter upper end of the torque tube displayed a 0.1-inch-wide and 0.028-inch-deep wear band. The band ran completely around the torque tube at the location corresponding to the upper end of the acme screw.

1.16.2.4 Lower Mechanical Stop and Corresponding Spline Area on the Acme Screw

The lower mechanical stop of the acme screw assembly was recovered with a section of the vertical stabilizer that had been separated from the jackscrew assembly and related structure. The lower mechanical stop was covered by dark, partially dried grease, and the stop bolt and nut were in place.

Numerous dents and mechanical marks were found on the upper and lower surfaces of the lower mechanical stop. Mechanical damage was also found on the lower mechanical stop's internal splines and on the clamp bolt. The lower mechanical stop's clamp bolt and nut were found in place.

Optical microscopy, scanning electron microscopy (SEM), and x-ray energy dispersive spectroscopy (EDS) examinations identified multiple damage features on the lower mechanical stop's upper surface, consistent with contact with the lower end of the acme nut and the nut's lower stop lug that protrudes below the main body of the nut. The features ranged from light marks in the primer paint to severe dents and substantial deformation of the metal. The damage surface was consistent with both clockwise (rotational) scoring and substantial deformation and rotational sliding contact (in both directions) with the acme nut.[134]

Rotational stripping was observed in circumferential bands at the upper and lower ends of the internal spline teeth in the lower mechanical stop.[135] The middle area of the

---

[133] For more information on torque tube testing, see section 1.16.5.

[134] In an intact jackscrew assembly, clockwise rotation of the acme screw results in movement of the horizontal stabilizer toward airplane-nose-down pitch, and counter-clockwise rotation of the acme screw results in movement of the horizontal stabilizer toward airplane-nose-up pitch.

splines, which was about 1/4 inch wide, was damaged but not stripped at any circumferential location. Circumferential scoring marks associated with the stripped areas of the spline teeth were on a plane slightly offset from directly circumferential. The stripping pattern and offset circumferential marks were consistent with the lower mechanical stop being at two or more skewed angles to the acme screw's splines during stripping. Most spline tooth peaks were found smeared in a counter-clockwise direction, consistent with a clockwise rotation of the acme screw (as viewed from above). However, many splines were damaged consistent with acme screw rotation in both directions.

Imprint marks created by upward motion of the acme screw spline teeth through the lower mechanical stop were observed across the peaks of the stop's spline teeth that were not stripped (the middle portion of the splines). The orientation of the marks was consistent with the lower mechanical stop being skewed (at an angle) to the splines of the acme screw as the screw pulled upward through the stop.

Damage along the length of the portion of the shank of the lower mechanical stop clamp bolt that was exposed to the interior of the stop was consistent with rotational contact with acme screw spline teeth. The exposed area of the bolt shank also had imprint marks oriented transverse to the length of the bolt, consistent with upward motion of the acme screw relative to the lower mechanical stop. These transverse marks on the bolt shank were superimposed upon the rotational damage, consistent with this damage being created, at least in part, after the rotational damage of the lower mechanical stop spline teeth.

1.16.2.5 Acme Nut

The acme nut was recovered attached to vertical stabilizer wreckage in its normal position but separate from the acme screw and the remainder of the jackscrew assembly. On-scene visual inspection of the thread area on the inside diameter of the acme nut showed a relatively smooth, flat surface, with only small ridges of the acme thread remaining. (Figure 15 shows an overall view of the recovered acme nut assembly. Figure 16 shows a closer view of the interior of the acme nut.) In addition, a pattern of dry, black deposits also spiraled around the diameter, following the raised ridges. The inside surface also contained random areas of white deposits. No evidence of grease was found inside the acme nut during on-scene visual and tactile inspections or during high-magnification examination in the laboratory. Various quantities of reddish-brown and black grease were found on the exterior of the acme nut and gimbal ring, and areas of

---

[135] Stripping was considered present in any region where individual spline teeth were indistinguishable from each other.

green-tinted, black and red grease deposits were found on the surrounding airplane structure.

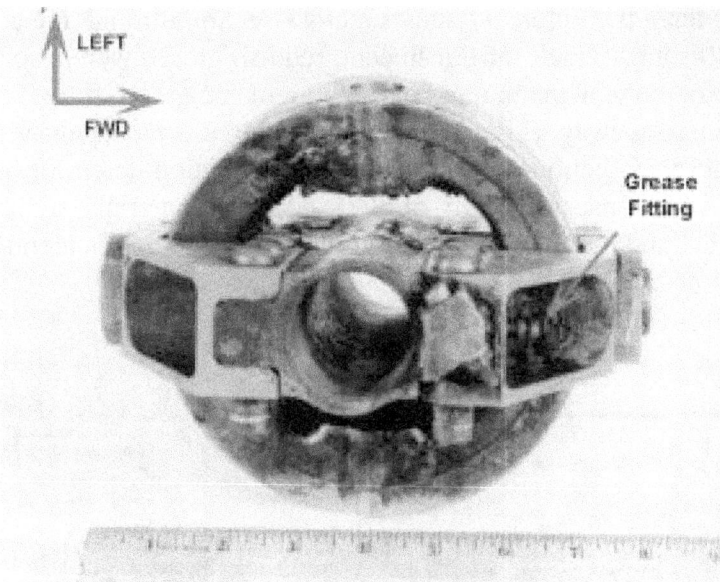

Figure 15. A photograph of an overall view of the recovered acme nut assembly.

Figure 16. A photograph of a closer view of the interior of the acme nut.

The acme nut has a grease fitting on the forward, upper side of the nut through which grease can be applied to the acme nut threads. A solid, black, claylike, dry residue coated with white, powdery material was found in the internal counterbore of the acme nut's grease passageway where it intersected the inside surface of the nut, as well as in the smaller diameter grease passageway behind the counterbore. When the grease fitting was removed to allow collection of material samples from the grease passageway, translucent, reddish grease consistent with the appearance of Mobilgrease 28 was observed on the fitting's external threads. (Figure 17 shows a cross-section through the acme nut's grease passageway and fitting.) Traces of translucent, reddish grease were also found inside the grease passageway in the acme nut and in the bore of the grease fitting, a section distinct from the grease passageway. (Figure 18 shows the grease passageway and counterbore after the acme nut was sectioned.) Microscopic examination determined that the sample from the counterbore consisted mostly of black powdery material, larger chunks of black and white material, and particles of copper-colored metallic debris identified as acme nut material.

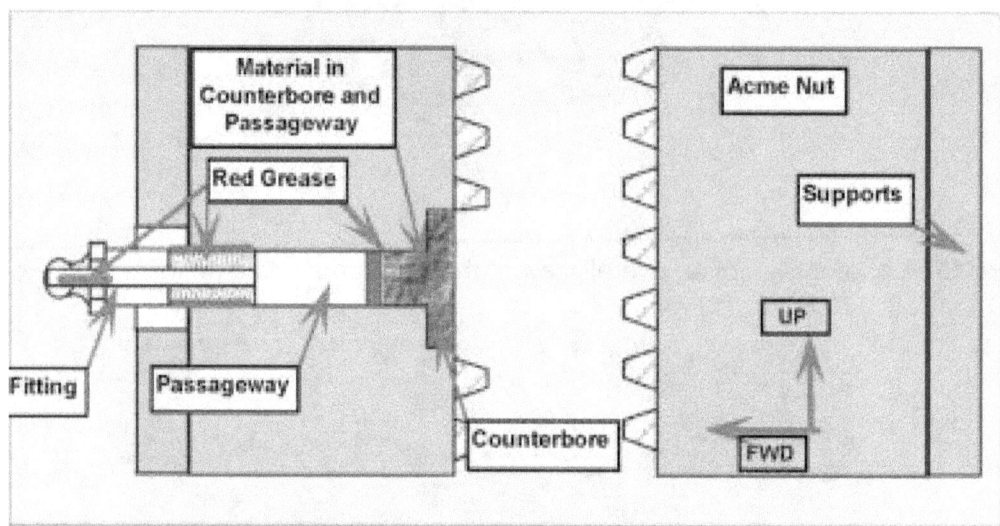

Figure 17. A diagram of a cross-section through the acme nut's grease passageway and fitting. Red areas denote the locations where red grease was found, and gray areas denote the location where the dry residue was found.

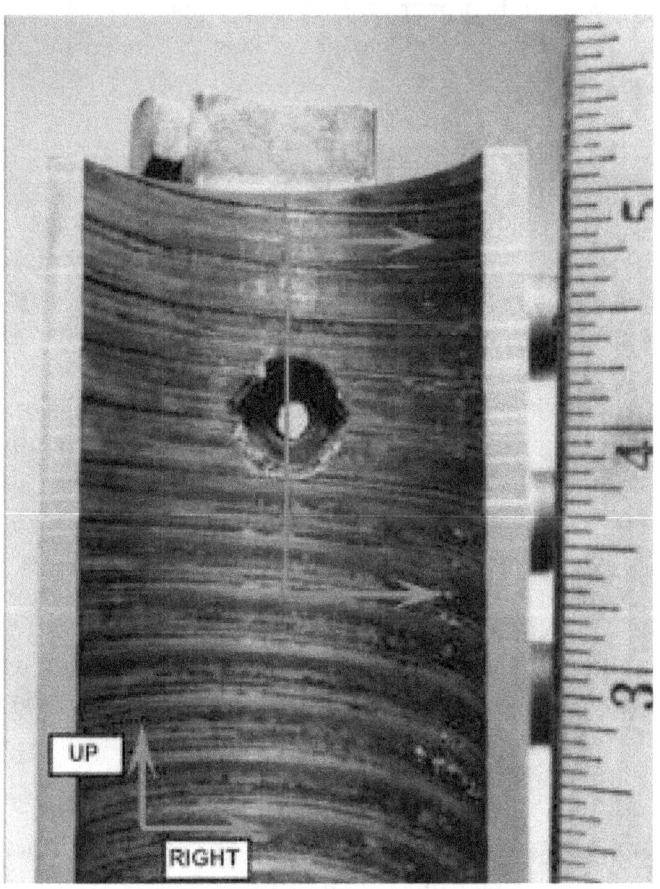

Figure 18. A photograph of the grease passageway and counterbore after the acme nut was sectioned.

Magnified visual inspections of the acme nut's interior after the nut was longitudinally sectioned confirmed the presence of the spiraling pattern consistent with the original location of the internal nut threads. For comparison with the accident acme nut interior, an exemplar nut was similarly sectioned.[136] The spiral patterns comprised a small raised ridge followed by a lustrous depression with spiral machining or wear marks transitioning into a dark band of surface deposits that rose to the next ridge. Under higher magnifications, longitudinal smearing and shear fractures were identified on the raised edges. The smearing and shearing directions were consistent with upward movement of the acme screw threads in relation to the nut body. Material was observed folded upward and smeared over adjacent surfaces along most of the ridges. Typical ridges were between 0.013 and 0.020 inch wide.

Most of the markings observed on the acme nut's interior surface spiraled around the bore in the same manner as the original thread pattern. However, some circumferential and longitudinal marks were observed in isolated locations overlaying the spiral marks. Two circumferentially oriented bands of scratches were observed partially around the

---

[136] The exemplar nut, which was removed during the overhaul of a jackscrew assembly, had a reported 0.0036-inch end play measurement at the time of overhaul.

inner diameter near the top of the acme nut on the forward side. Several longitudinal scuff marks were also observed on the upper aft side of the acme nut's interior above the grease passageway. The scuff marks appeared to have been produced by the contacting object moving vertically upward.

1.16.2.6 Acme Nut Thread Remnant Examination

Twelve acme nut thread remnants[137] were found wrapped around the central part of the acme screw when it was recovered. Most of the thread remnants were found spiraled around the acme screw oriented primarily along the path of the screw threads, although some were found tangled together and bent in multiple directions. The thread remnants were clustered into two closely spaced groups extending about 6 to 8 inches along the acme screw. The thread remnants were copper colored with a greenish tint, consistent with copper-alloy oxidation. After recovery, each thread remnant was numbered, and its width, maximum thickness, estimated length, and number of 360° turns were measured. The accumulated length of the thread remnants was equivalent to about 75 percent of the total length of the original threads.

High-magnification optical examinations revealed that all of the thread remnants were roughly rectangular in shape with a mostly flat upper side and a multibeveled lower side. The outer diameter's edges were fractured, corresponding to the approximate major (outer) diameter of the acme nut. Examination also revealed that some of the thread remnants had corrosion pitting.

A representative sample of a thread remnant (No. 4) was further examined by SEM and EDS. EDS examination determined that thread remnant No. 4 was composed of an aluminum-nickel-bronze alloy, consistent with acme nut material specified in the manufacturer's engineering drawing (unified numbering system [UNS] alloy C95500).[138] High-magnification SEM examination revealed smearing deformation and shearing dimples along thread remnant No. 4's outer edge fracture surface. The orientation of the shear dimples was consistent with thread remnant No. 4 moving upward relative to the body of the acme nut. The following measurements were made at one of the fractured ends of thread remnant No. 4: the shear fracture on the outer edge measured 0.0131 inch, the nearby minimum section thickness measured 0.0076 inch, and the maximum section thickness at the central ridge measured 0.0184 inch.

Additionally, two metallographic cross-sections were cut radially through another representative remnant (No. 6). The metallographic sections showed that thread remnant No. 6's flat face exhibited a slight, concave curvature, which was also identified on thread remnant No. 4. The following measurements were made on one of the metallographic sections of thread remnant No. 6: the shear fracture on the outer edge measured 0.0174

---

[137] One of these acme screw thread remnants was removed by Safety Board investigators on scene.

[138] UNS alloy C95500 is a copper alloy containing aluminum, nickel, iron, and manganese. The engineering drawing specified an aluminum-nickel-bronze alloy per QQ-B-671, type 2, class 4 Cond HT. This designation has been superseded with UNS alloy C95500.

inch, the nearby minimum section thickness measured 0.0099 inch, and the maximum section thickness at the central ridge measured 0.0180 inch.

1.16.3 Shear Load Capability of the Acme Nut Threads

The ultimate shear load capability of the acme nut threads was approximated by assuming that the thread root cross-section was uniformly at the material's ultimate shear stress. This calculation was performed for three thread thicknesses: 0.15-inch-thick threads, as measured on newly manufactured threads; 0.0131-inch-thick threads, as measured on thread remnant No. 4; and 0.0174-inch-thick threads, as measured on the metallographic section of thread remnant No. 6. The calculations were based on the specified minimum ultimate strength (Ftu) for the nut material and assumed a shear strength equal to 77 percent Ftu. Table 4 shows the results of the shear load calculations.[139]

Table 4. Shear load calculations for three acme nut thread thicknesses.

	Shear Load Calculations of the Acme Nut Threads		
	Shear load for 0.15-inch-thick thread (newly manufactured)	Shear load for .0131-inch-thick thread (as measured on thread remnant No. 4)	Shear load for .0174-inch-thick thread (as measured on the metallographic section of thread remnant No. 6)
Load for 32 threads	2,200,999 pounds	191,263 pounds	249,782 pounds
Load for 1 thread	68,781 pounds	5,977 pounds	6,806 pounds

1.16.4 Studies of Thread Stress and Deformation

In connection with this investigation, the Safety Board asked Boeing to conduct a study to estimate the maximum wear that the acme nut threads could sustain and still maintain their ultimate load capability. Additionally, the Board, in collaboration with the State University of New York, Stony Brook, New York, initiated a study of the stresses and deformations in the acme nut. Both studies utilized the ABAQUS[140] finite element analysis (FEA) code and axisymmetric models of the acme screw and nut to address thread-by-thread distribution of the load transfer from the screw to the nut. The Boeing study also used an axisymmetric model to evaluate the stresses in a loaded thread section of the same geometry as one of the remnants recovered from the accident airplane. The Safety Board/Stony Brook study used a three-dimensional FEA to examine the change in deformation and stress in a single thread revolution as a function of acme nut thread wear.

---

[139] For more information about these calculations, see the Materials Laboratory Group Chairman's Factual Report in the public docket for this accident.

[140] Hibbitt, Karlsson & Sorensen, Inc. *ABAQUS/Standard User's Manual*. Version 6.2. Pawtucket, Rhode Island. 2001.

Both studies indicated that the load transfer among the 32 thread revolutions was not uniform along the length of the acme nut, with thread revolutions near the top and bottom of the nut receiving greater loads than those near the center of the nut. Further, the studies showed that the load distribution varied depending on specific temperature and load conditions. The Boeing study concluded that the maximum load carried by any individual thread revolution was approximately 30 percent of the applied load, whereas the Safety Board/Stony Brook study concluded that the maximum load carried by any individual thread revolution was approximately 23 percent of the applied load. On the basis of the results of the thread remnant stress analysis, Boeing concluded that the only way to develop the large plastic bending observed in the thread remnant was for the thread loading to be concentrated near the tip of the thread rather than near the root.

The Safety Board/Stony Brook study examined the acme screw-to-nut contact pressure distribution on a single nut thread revolution and determined that the pressure distribution was nonuniform across the thread surface (radial to the nut), with a small region near the thread root (major diameter) having a much higher pressure than the remainder of the contact surface. When material removal was modeled in the region of highest stress, the stress pattern stayed the same with the high-contact pressure region moving radially inward across the thread from the material removal region to the adjacent material on the remaining thread surface. The Safety Board/Stony Brook study concluded that the wear mechanism consisted of material removal proceeding on a layer-by-layer removal process along the thread surface and that, within each layer, wear progressed from the root to the tip (major diameter to minor [inner] diameter).

The Safety Board/Stony Brook study ran a series of FEAs to evaluate the effect of the sequential wear process on the stress and deformation in acme nut threads. The study found that the stresses in the acme nut remained low compared to the yield stress[141] of the aluminum-bronze acme nut material, even at wear levels exceeding the maximum amount allowed for an acme screw to remain in service. At an axial wear level of approximately 0.090 inch, the maximum stress was beginning to approach the yield stress under a load of 2,000 pounds per thread revolution. When the axial wear level reached 0.09265 inch and approximately 50 percent of the radial thread surface had worn, the bending deflection of the thread tip began to increase rapidly with load increase. When approximately 75 percent of the radial thread surface was removed, the bending reduced the available contact surface to a value too small for the FEA to provide valid data. The study concluded that, at this point, the acme screw thread would begin to slide axially across the thread surface rather than slide during rotation.

1.16.5 Torque Tube Testing

Under the Safety Board's supervision, Boeing conducted static load tests on torque tubes. Torque tube assemblies from MD-80 airplanes other than the accident airplane were

---

[141] Yield stress is the largest value of stress for which all deformation within the material is recoverable when the material is unloaded.

tested to failure in tension to generate the sound signature[142] of the predicted torque tube fracture and to validate and generate data for a subsequent FEA.

To conduct the static load tests, the threads on two acme nuts were machined away to their roots (the major diameter) so that the acme screw could move freely through the nut vertically (without rotating). The tests were conducted in a computer-controlled hydraulic load frame by pulling the acme screw upward through the altered nuts until the torque tube fractured.

In the first static load test, the lower mechanical stop on the acme screw contacted the acme nut in a manner that did not create any bending loads in the torque tube. This produced pure tension loading in the torque tube similar to loading that occurs during normal flight conditions. The first static load test produced a tension failure of the torque tube at a load of 74,400 pounds.[143]

In the second static load test, the lower mechanical stop on the acme screw contacted the acme nut in a manner that caused the entire load to be applied to one side of the stop in an offset manner. This type of load was similar to that indicated by lower mechanical stop damage on the accident airplane; however, the load was applied on top of the stop lug, creating a moment arm slightly greater than that indicated on the accident lower mechanical stop. This loading condition produced a combination of tension and bending in the torque tube. The second static load test produced a tension failure of the torque tube at a load of 25,576 pounds.[144]

For the FEA, Boeing generated a numerical model for the geometry and material parameters of the accident acme screw, torque tube, lower mechanical stop and clamp bolt, washer, and lower assembly nut. This model was used to analytically calculate the static offset load capability of the accident torque tube using the loads derived from the two static load tests. The Boeing model predicted that the accident torque tube fractured at a load of 25,500 pounds.

Neither of the two static load test conditions accounted for fatigue damage to the torque tube (similar to damage found on the accident torque tube). Low-cycle fatigue testing of small diameter torque tube material specimens was performed to estimate the stress conditions that would generate a similar low-cycle fatigue fracture surface. To test the torque tube material, tensile specimens were machined from torque tubes; threads were added to the center portions of the specimens (simulating the threads at the lower end of the torque tube); and constant-amplitude cyclic loading at calculated stress levels[145] of 140,000, 166,000, and 185,000 pounds per square inch (psi)[146] were applied to the

---

[142] For more information on the Safety Board's sound spectrum study, see section 1.11.1.1.

[143] When the airplane is subjected to the ultimate design loads specified for certification, the torque tube sustains a 57,000-pound load.

[144] According to the Boeing torque tube test report summary, "the fracture load of just over 25,000 [pounds]…is based on a loading configuration that provides for the largest offset moment arm with the nut stop resting directly upon the lower rotational stop." Boeing's summary noted that the accident airplane's "lower rotational stop exhibits evidence that the orientation of the two stops provides a smaller moment arm" and thus had a higher predicted fracture load.

individual test specimens until a fracture was produced. Low-cycle fatigue fractures were created under all three of the test conditions. The fracture surfaces on the accident torque tube were more similar to those produced when 140,000- and 166,000-psi cyclic loads were applied than to those produced when a 185,000-psi cyclic load was applied. The 140,000-psi specimen fractured after application of 1,549 load cycles, the 166,000-psi specimen fractured after application of 389 load cycles, and the 185,000-psi specimen fractured after application of 100 load cycles.

1.16.6 Additional Safety Board Jackscrew Assembly Examinations

1.16.6.1 February 2000 Post-AD 2000-03-51 Jackscrew Assembly Examinations

The Safety Board examined eight jackscrew assemblies removed from airplanes shortly after the issuance of FAA AD 2000-03-51 in February 2000,[147] including two assemblies from Alaska Airlines' airplanes.[148] The jackscrew assemblies were removed in accordance with AD 2000-03-51 for various reasons, including wear and the presence of metal shavings. In addition, the Board examined five Alaska Airlines acme nuts that had been removed from jackscrew assemblies that were being overhauled.

The Safety Board performed a loaded acme nut end play test on all eight of the removed jackscrew assemblies by mounting each jackscrew assembly (in as-received condition) in a frame and attaching the gimbal ring to a moment arm. During the tests, the acme nut was loaded by applying 180 pounds in both directions (up and down). Movement of the acme nut on the screw was measured using a dial indicator and recorded as end play. End play measurements were made with the acme nut positioned in three locations: near the center of the acme screw, near the lower mechanical stop, and near the upper mechanical stop. The free play was also measured at the middle position during the loaded end play test.

After lubricating the middle section of the acme screw with Aeroshell 33 by hand and injecting it through the acme nut's grease fitting (the original grease was not removed), end play measurements were repeated on each jackscrew assembly with the nut in the middle position. The procedures were repeated again after removing by hand all of the grease from the middle section of the acme screw. The test data show that the as-received end play measurements changed slightly in both directions for both the regreased and new grease conditions.

---

[145] The stress level was calculated by dividing the applied load by the measured cross-sectional area in the threaded region of the test specimens.

[146] These values were used to ensure that testing occurred in the high-stress region of the torque tube and produced a fracture in a short timeframe. The values were first estimates of the possible stress range that the torque tube might have experienced and were not intended to reproduce actual conditions.

[147] For more information about postaccident ADs, see section 1.18.1.

[148] These jackscrew assemblies were DCA 3000 from N982AS and DCA 3008 from N981AS.

All eight of the removed jackscrew assemblies were disassembled to separate the gimbal ring from the acme nut, the acme nut from the acme screw, the acme screw from the torque tube, and the torque tube from the spherical bearing. Individual items of each jackscrew assembly were cleaned and visually examined. Visual inspections of the disassembled acme nuts from the eight removed jackscrew assemblies and the other five Alaska Airlines acme nuts identified sharp edges, or burrs, at the minor diameter of all of the threads on all of the nuts. The burrs on all of the acme nuts extended inward from the lower flanks of the threads. Some burrs were partially disconnected, forming long, thin metal slivers. The lower flanks and major diameters of the threads were bright and shiny with a finely scratched and worn appearance. The scratches followed the spiral threads and had obliterated the original surface finish.

One of the acme nuts from an Alaska Airlines jackscrew assembly that was being overhauled was cut along its longitudinal axis to inspect its thread surfaces. Burrs and slivers were identified along the entire length of the cut. In many areas, slivers were separating from the threads in thin, wide flakes. The lower flanks and the major diameter of the threads were worn along the entire acme nut. Almost no wear marks were identified on the minor diameters or on the upper flanks.

Most of the acme screws from the disassembled assemblies were very similar in wear characteristics, showing only polishing of the lower thread flanks and major diameter. However, the acme screw from one of the removed jackscrew assemblies (S/N DL-96) showed much heavier wear with steps worn into both the upper and lower flanks of the threads. During the disassembly of DL-96, gritty, gray-colored material was noted on the upper surfaces of the gearbox support plates.

The torque tubes from all of the jackscrew assemblies were undamaged, except for DCA 3000 from N982AS. The upper end of this torque tube displayed a .0065-inch-deep wear band all around the large diameter section.

1.16.6.2 Examinations of Hawaiian Airlines' Jackscrew Assemblies

On February 17, 2000, a jackscrew assembly, S/N DCA 110, was removed from a Hawaiian Airlines DC-9 (N601AP) after an on-wing end play measurement of 0.043 inch was recorded.[149] The assembly had accumulated 2,582 flight hours (6,377 flight cycles) since a 1997 overhaul.[150] A bench end play check conducted by Integrated Aerospace[151] after the jackscrew assembly was removed and cleaned yielded an end play measurement of 0.048 inch. The acme nut and screw were forwarded to the Safety Board's Materials Laboratory for examination.

---

[149] Before the flight 261 accident, Hawaiian Airlines' lubrication interval for the jackscrew assembly was every 1,000 flight hours, and its end play check interval was every 3,000 flight hours.

[150] After the 1997 overhaul, the DCA 110 jackscrew assembly had a recorded end play measurement of 0.010 inch. The jackscrew assembly was sent to Hawaiian Airlines, where it was stored until its installation on N601AP on June 22, 1998.

[151] Integrated Aerospace, formerly Trig Aerospace, Inc., is the only original manufacturer and supplier of new jackscrew assemblies.

Metallurgical examination of the acme screw identified visible wear steps in both the upper and lower flanks of the threads. No other mechanical or corrosion damage was noted on the acme screw. Examination of the acme nut showed that both flanks of the nut threads were lustrous with circumferential marks consistent with wear. However, no burr was noted at the minor diameter as seen on other acme nuts previously examined. The major diameter of the threads (the root area) was also worn and lustrous with a fine pattern of parallel scratches on both the upper and lower flank surfaces. The parallel scratch pattern ran circumferentially around the thread flanks. The major diameter surfaces also showed heavy localized wear that enlarged the diameter in some areas of the acme nut. The diametrical wear was determined to be nonuniform in both the circumferential and vertical directions, consistent with an angular misalignment of the acme screw axis to the acme nut axis with the upper end of the screw displaced to the right and the lower end to the left. The calculated wear rate of this jackscrew assembly (using the 0.048-inch n end play measurement obtained during the overhaul bench check and accounting for the 0.010-inch end play measurement obtained when the jackscrew assembly was first installed) was 0.0147 inch per 1,000 flight hours.

On June 18, 2001, another jackscrew assembly, also partially stamped with S/N DCA 110,[152] was removed from the same airplane after an on-wing end play measurement of 0.036 inch. The acme screw and nut had accumulated 2,514 flight hours (6,071 flight cycles). Metallurgical examination at the Safety Board's Materials Laboratory identified polishing on the acme screw's upper thread flanks and major diameter in the central part of the screw. No steps in the thread flanks were identified. The lower flanks of the acme nut threads were worn with a circular pattern, and the shape of the acme screw thread major diameter chamfer was worn into the nut threads. Examination of the acme nut's profile and interior surfaces after sectioning identified small burrs on all threads extending from the lower thread flank into the minor diameter area. The major diameter surfaces displayed some localized contact damage, but the diameter was not enlarged at any location. The calculated wear rate of this jackscrew assembly (assuming a starting end play measurement of 0.006 inch after overhaul and a final bench-check end play measurement of 0.027 inch) was 0.0084 inch per 1,000 flight hours.

The Safety Board's examination also revealed evidence of pink-colored grit material inside the counterbore of the acme nut grease passageway in the jackscrew assembly removed in February 2000. A Hawaiian Airlines maintenance work card dated December 4, 1999, stated, "Grit blast accumulated in vertical stab as seen thru access panel 6401 and panel 3703 (left lower banana fairing). Area to be inspected after cleaning." The corrective action noted on the card was the following: "Cleaned area. Inspected area. No discrepancies noted." According to Hawaiian Airlines personnel, the

---

[152] Although S/N DCA 110 was stamped on the gearbox support assembly, the S/N stamped on the acme screw and nut was D 3351, indicating that the acme screw and nut from the first DCA 110 assembly were replaced by a new acme screw and nut. Hawaiian Airlines personnel stated that they kept the DCA 110 gearbox support assembly, torque tube, and stop nut with N601AP (after sending the acme screw and nut for overhaul) when they installed the newly acquired D 3351 acme nut and screw in February 2000. Records indicated that the D 3351 assembly was manufactured in 1998 and delivered to Boeing with an end play measurement of 0.006 inch.

area noted on the work card was located inside the vertical stabilizer where the jackscrew assembly is installed.

Several grease samples taken from the original acme nut and screw of jackscrew assembly DCA 110 and from the area of the assembly near the inside of the vertical stabilizer contained translucent, pink particles, consistent with the fine grains of a sandlike material. EDS chemical analysis indicated that the grains were composed of iron, silicon, aluminum, calcium, oxygen, and magnesium. Examination of grease samples taken from the D 3351 assembly also revealed pink particles, consistent with those found in the grease sample taken from DCA 110.

An on-site inspection of Hawaiian Airlines' maintenance facilities in August 2001 revealed that it had been using either Australian garnet or glass beads to remove corrosion from airplane structure. Examinations of a sample of the garnet material found that it visually and chemically matched the pink, sandlike material found in the grease of both jackscrew assemblies removed from Hawaiian Airlines.

1.16.6.3 Examinations of Alaska Airlines Reports of Acme Screw "Wobble"

On November 26, 2000, Alaska Airlines maintenance personnel observed the acme screw "wobble" while operating the horizontal stabilizer trim system on N965AS. According to the personnel, the wobble appeared as a slight rocking motion of the acme nut as the acme screw was rotated within it. Alaska Airlines removed the jackscrew assembly from N965AS and notified the Safety Board about the wobble.

On February 21, 2001, Safety Board investigators observed the jackscrew assembly as it was operated on an Integrated Aerospace fixture and a wobble was noted. The end play of the wobbling acme screw was checked several times (before disassembly) while installed on the Integrated Aerospace fixture. The acme screw was rotated before each end play check to identify any variance in end play caused by wobble-induced acme nut and screw position changes. No differences were noted, and the end play measured 0.018 inch consistently.

After the jackscrew assembly was disassembled, the technician who removed the lower torque tube nut indicated that the torque nut was looser than what he considered to be typical.[153] A wear band was noted on the large diameter section near the top portion of the torque tube. The depth of the wear band was not uniform around the periphery of the groove and measured 0.035 inch at its deepest point. According to Boeing, a wear band frequently occurs in this area because the clearances are small and, at times, the torque tube contacts the inside wall of the acme screw. The Boeing DC-9 Overhaul Maintenance Manual requires a steel-sleeve repair of the wear band during overhaul if the wear band is deeper than 0.025 inch.

---

[153] The technician used a wrench to remove the torque nut. No measuring device was used to measure the torque required to loosen the nut. The technician's comments were based on his personal experience of feel.

In a March 27, 2001, letter, Alaska Airlines asked the Safety Board to investigate the wear band. Alaska Airlines noted that the torque tubes from N982AS and the accident airplane had similar wear bands.[154] Alaska Airlines also expressed concern that the wobble could produce erroneous end play measurements and/or excessive acme nut wear. The Board collected data from Integrated Aerospace and Boeing to determine if a relationship existed between acme nut wear rate and torque tube wear band depth. The collected data included information about 161 jackscrew assemblies examined by Integrated Aerospace in 2000. Of these assemblies, only 77 had torque tubes. Of the 77 jackscrew assemblies with torque tubes, only 49 had a recorded bench-check end play measurement, a torque tube wear band measurement, and a valid flight-hour total. A Board examination found no statistically significant difference regarding acme nut wear between the torque tubes with no (or a slight) wear band and the torque tubes with deeper wear bands.

1.16.7 Chemical and Microscopic Analyses of Grease Residues from the Accident Jackscrew Assembly

The Safety Board contracted the U.S. Navy's Air System Command's Aerospace Materials Laboratory, Patuxent River, Maryland,[155] to conduct chemical and microscopic analyses of eight grease residue samples taken from the accident jackscrew assembly. An organic analysis method, Fourier Transform Infrared Spectroscopy, was used to compare the grease residue samples taken from the accident jackscrew assembly to control samples of Aeroshell 33 and Mobilgrease 28. The spectra resulting from the residue samples were weak, which was attributed to the degraded quality of the grease. None of the grease residue samples could be matched perfectly to either Aeroshell 33 or Mobilgrease 28. However, the resulting analysis indicated that most of the residue samples displayed infrared spectra consistent with the presence of both Aeroshell 33 and Mobilgrease 28. On the basis of a comparison of the residue samples to the control samples, the Naval laboratory concluded that both greases were likely present on the accident jackscrew assembly. An independent review of this work conducted by a private consultant commissioned by the Board also concluded that the residue samples had infrared signatures consistent with the presence of both Aeroshell 33 and Mobilgrease 28.

According to the Naval laboratory report, the microscopic analysis of the residue samples found that the "sample grease was heavily loaded with particles[156] originating from…aluminum bronze" and that "these particles were consistent with wear debris." The

---

[154] Metallurgical examination determined that the wear band on the torque tube from jackscrew assembly DCA 3000 was 0.0065 inch deep and that the wear band on the accident torque tube was 0.028 inch deep.

[155] The U.S. Navy's Aerospace Materials Laboratory is the Department of Defense's (DoD) steward of the MIL PRF specifications for both Aeroshell 33 and Mobilgrease 28 and qualifies greases to these specifications.

[156] The Naval laboratory report stated that "a couple of larger metal particles were extracted from the grease with tweezers and examined individually. One was a large flake and the other was a long shaving. Both were aluminum bronze, equivalent in composition to the smaller particles."

Naval report stated that "no other particle contamination was found other than that attributed to the environment of the crash site and the recovery operation."

1.16.8 Additional Grease Testing, Experiments, and Analysis

1.16.8.1 Standardized Grease Testing

The Safety Board also contracted the U.S. Navy's Aerospace Materials Laboratory to conduct a series of standardized grease tests. Pure quantities of Aeroshell 33 and Mobilgrease 28, as well as several mixture ratios of each grease, were tested using several American Society for Testing and Materials standard test methods for lubricating greases. These tests examined the greases' firmness (when new and after being mixed), storage stability, and oil bleedout caused by elevated temperature and elapsed time. The results of these tests indicated that, in general, the physical properties of Aeroshell 33 and Mobilgrease 28 did not change significantly when mixed; however, the change in the physical properties of the two greases noted in 90/10 and 10/90 mixture ratios exceeded that of the standard for compatibility of mixed greases. The Naval laboratory report concluded that the change in properties was insignificant and would not significantly affect the greases' performance.

The Naval laboratory also conducted seawater immersion tests on control samples of Aeroshell 33 and Mobilgrease 28 and on a 50/50 mixture of the two grease types. The Naval report concluded that, after being immersed in seawater for 2 weeks, the three test samples "exhibited only slight color changes, with no other adverse effects noted." The report concluded that "deterioration of the aircraft grease sample [from the accident airplane] was not due to exposure to seawater for two weeks following the incident."[157]

1.16.8.2 Experiments and Analysis on the Surface Chemistry of Acme Nut Material When Exposed to Various Greases or Grease Mixtures

The Safety Board contracted the U.S. Navy's Aerospace Materials Laboratory and Science Applications International Corporation, San Diego, California, to conduct a series of experiments to examine the surface chemistry of acme nut material when exposed to Mobilgrease 28, Aeroshell 33, and mixtures of the two greases under various conditions that can occur during flight, including exposure to ambient and elevated temperatures and fluid contamination (water, deicing fluid, and anti-icing fluid).

During the experiments, it was found that under certain circumstances Aeroshell 33 would produce a visible brownish discoloration to small, localized regions of the acme nut material. The surface chemistry of the discolored regions was analyzed using x-ray photoelectric spectroscopy. The analysis indicated that the discolored region was

---

[157] In another example, the Safety Board observed that an acme screw recovered from a China Northern MD-80 airplane that crashed into a salt water bay near Dalian, China, on May 7, 2002, still had a coating of grease clearly visible on it after it had been immersed in seawater for 5 days. According to the Civil Aviation Administration of China, the last scheduled lubrication of the jackscrew assembly was during an A check that took place March 27 through April 2, 2002, and the specified grease was Aeroshell 5 (MIL-G-3545C).

caused by additives in the Aeroshell 33 grease chemically reacting with each other and depositing on the surface of the acme nut material, which is an intended chemical reaction between antiwear additives in the grease. Science Applications International Corporation concluded that this was not a corrosion process and, therefore, would not corrode the acme nut material.

1.16.8.3 Wear Testing Under Various Grease Conditions

The Safety Board contracted the Tribological Materials Laboratory at Rensselaer Polytechnic Institute, Troy, New York, and the tribology division of Battelle Memorial Institute, Columbus, Ohio, to conduct a series of wear tests on grease-lubricated, aluminum-bronze and steel couples in sliding wear. The test program concentrated on the American Society for Testing and Materials D3704 Block-on-Ring wear test method, which involves a steel ring sliding in reciprocating motion upon an aluminum bronze block. Contact pressure ranges and sliding speeds used in these tests reflected those experienced in an operational jackscrew under flight and static loads. Additionally, test specimens were fabricated from actual jackscrew assembly steel and aluminum-bronze components.

Because Mobilgrease 28 has been demonstrated to effectively lubricate the jackscrew assembly, the wear test program examined the lubricating effectiveness of Aeroshell 33 relative to Mobilgrease 28. The wear tests were run under the following conditions: (1) with a mixture of Aeroshell 33 and Mobilgrease 28; (2) with greases contaminated with water, deicing fluid, and salt water; (3) with aged and wear-debris laden grease; (4) with greases exposed to subzero temperatures; and (5) with a complete lack of lubrication. Over 50 wear tests were performed. During the wear tests, the wear rates were continuously monitored so that the steady-state wear behavior, which governs the long-term jackscrew assembly wear, could be identified separately from the transient wear behavior experienced during the break-in period that began each test.

The results of the wear tests showed that, in all ranges of contact pressure, lower wear rates were generated when Aeroshell 33 was used than when Mobilgrease 28 was used. Additionally, the effects of contamination, aging, and subzero temperature exposure were determined to have little effect on the lubricating effectiveness of Aeroshell 33. Only the tests run without any lubrication generated wear rates significantly higher than those generated when Mobilgrease 28 was used. In these cases, wear rates were seen to transition up to 10 times higher than those with equivalent conditions lubricated with Mobilgrease 28. Figure 19 shows a summary of the results of the high-load wear tests, including the contamination and low-temperature tests.

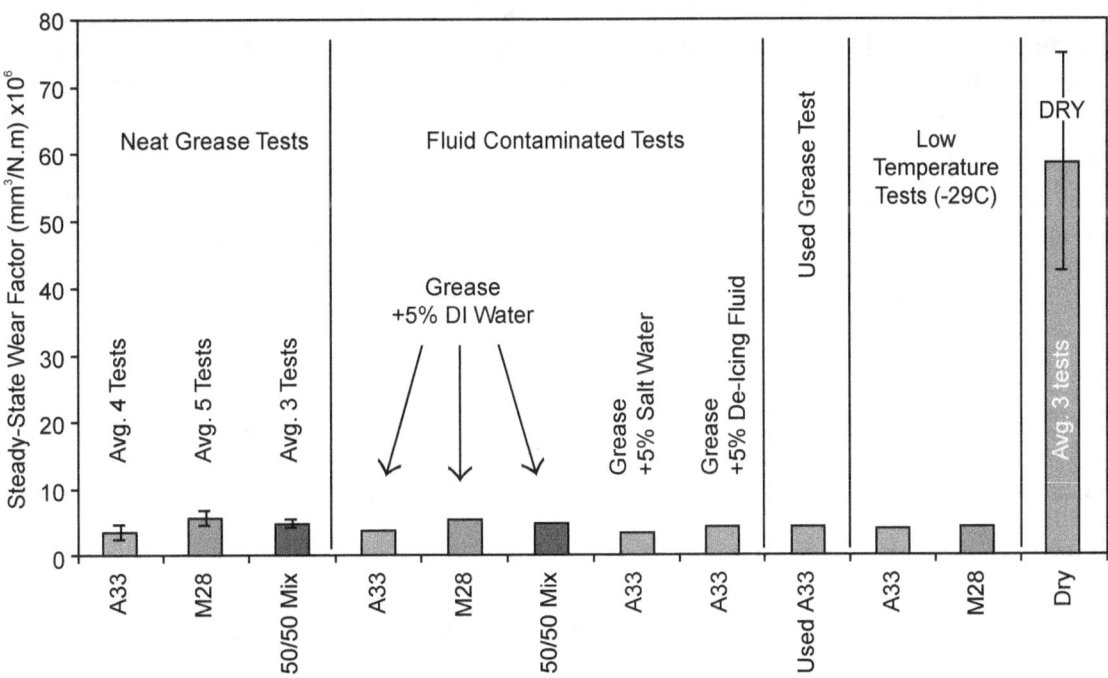

Figure 19. A summary of the results of the high-load wear tests, including the contamination and low-temperature tests.

1.17 Organizational and Management Information

1.17.1 Alaska Airlines, Inc.

Alaska Airlines, Inc., was organized in 1932 and incorporated in 1937 and, at the time of the accident, was one of two subsidiaries of Alaska Air Group, Inc., which was incorporated in 1985. At the time of the accident, Alaska Airlines employed about 10,000 people and served more than 40 domestic and international destinations (including 5 cities in Mexico) from hubs in Anchorage, Alaska; Seattle, Washington (the location of the company's headquarters); and Portland, Oregon. According to FAA documents, at the time of the accident, Alaska Airlines operated a fleet of 88 airplanes, including 49 737s and 39 MD-80 series airplanes.[158]

---

[158] According to a June 20, 2000, FAA special inspection report on Alaska Airlines, the airline was operating a fleet of 89 airplanes, of which, 55 were 737s and 34 were MD-80 series airplanes. The FAA special inspection report added that Alaska Airlines planned to add 9 737-700 and 10 737-900 series airplanes to its fleet over the next 2 years. For more information about the FAA special inspection report, see section 1.17.3.4.

1.17.2 Alaska Airlines Flight Crew Training

According to FAA and Alaska Airlines documents, Alaska Airlines conducted training for its MD-80 fleet under the Advanced Qualification Program, which was effective March 1, 1985. A Safety Board review of the airline's MD-80 training curriculum indicated that the primary and alternate horizontal trim system components were covered during the airplane-specific ground school. During the second segment of maneuver briefing (before the full-flight simulator session), pilots were asked to describe situations that might require use of the runaway stabilizer checklist and to perform the checklist.[159] Application and performance of the runaway stabilizer checklist was also required during the third and fourth segments of the curriculum.

Classroom materials for the MD-80 continuing qualification program, which were effective in 1999, included normal and abnormal procedures for flight control systems. Supplemental instructional material given to pilots contained a quiz that included questions addressing electrical failures and associated stabilizer inoperative indications.

1.17.2.1 Stabilizer Trim Check Procedures

The stabilizer trim check procedures listed in the Alaska Airlines MD-80 Flight Handbook included checking both the primary and alternate trim systems. Specifically, the "Before Start Expanded Procedures" included the following steps:

> Move PRIMARY MOTOR BRAKE switch to 'STOP.' Operate both Captain's control wheel trim switches in NOSE UP and NOSE DN [down] directions. **Observe stabilizer does not move and aural warning does not sound. Unless stabilizer is running away, NEVER PLACE THE PRIMARY MOTOR BRAKE SWITCH TO 'STOP' WITH THE STABILIZER IN MOTION. Damage will result.** Move the PRIMARY MOTOR BRAKE switch to NORM and verify guard is down. **The PRIMARY MOTOR BRAKE switch does not affect operation of the alternate stabilizer trim system.** While observing the stabilizer indicator, hold Captain's control wheel left trim switch to NOSE UP and then to NOSE DN momentarily.
>
> **CAUTION** – Do not hold single switch position longer than 3 seconds.
>
> **The stabilizer should not move and aural warning should not sound.** Repeat this check using Captain's control wheel right trim switch. Observe stabilizer does not move and aural warning does not sound.
>
> Operate both Captain's control wheel trim switches in NOSE UP and then NOSE DN directions and note corresponding direction of stabilizer indicator and aural warning. Operate First Officer's control wheel trim switches and check stabilizer indicator. **Moving Captain's and First Officer's control wheel trim switches in opposition directions will stop stabilizer movement. This is not a required test. If test is used, do not exceed 3-second limit.**

---

[159] For more information about the runaway stabilizer checklist, see section 1.17.2.2.

Operate both LONG TRIM HANDLES in the same direction and check for corresponding trim indicator movement, release handles to neutral. While observing horizontal trim indicator, momentarily operate each ALT LONG TRIM switch individually to NOSE UP and NOSE DN. Observe stabilizer does not move and aural warning does not sound. Operate both ALT LONG TRIM switches and check for direction of travel at slower rate. **To obtain maximum service life, do not maintain switch positions longer than necessary to check response. If there is no response, release switches and determine cause. Should either trim handle or any control switch bind, hesitate or stick when released to neutral, it should be corrected before flight. Any time LONG TRIM handles, control wheel trim switches, or ALT LONG TRIM (secondary) controls are used, they should be moved to their fully deflected switch/lever position to ensure both motor and brake circuits are actuated.**

The "Taxi Expanded Procedures" in the Flight Handbook outlined the following steps:

Rotate the FLAP thumbwheel, in the LONG TRIM Takeoff Position Display, until the computed TO (takeoff) FLAP value appears in the FLAP window.

Rotate the CG thumbwheel, in the LONG TRIM Takeoff Position Display, until the computed CG value appears in the CG window.

- when the FLAP and CG values are set for takeoff, the Stabilizer TAKEOFF CONDTN LONG TRIM Readout window will display the proper trim setting.
- using the Primary LONG TRIM Control Wheel Switches, or LONG TRIM Handles, set the LONG TRIM (White) Indicator opposite the Long Trim Takeoff Position (Green) Indicator.
- if the indicators are not within specified tolerance, after takeoff flaps have been selected, the aural warning will sound when the throttles are advanced for takeoff.

1.17.2.2 Runaway Stabilizer Checklist Procedures

The Alaska Airlines MD-80 Quick Reference Handbook (QRH) included the following checklist procedures to troubleshoot a runaway stabilizer:

PAGE 18   EMERGENCY   *Alaska Airlines*

RUNAWAY STABILIZER

NOTE: Runaway trim may be confirmed by
grasping the trim indicator.

IF THE INDICATOR HAS REACHED THE FULL NOSE UP
OR DOWN LIMIT, ATTEMPT TO RETRIM THE AIRCRAFT
WITH THE LONGITUDINAL TRIM "SUITCASE" HANDLES.

1. STABILIZER TRIM (RED GUARDED SWITCH) .... STOP

2. LONGITUDINAL TRIM INDICATOR .... OBSERVE

♦ | LONGITUDINAL TRIM INDICATOR STOPS | **OR** | LONGITUDINAL TRIM INDICATOR CONTINUES TO MOVE | ⇨

⬇

♦ 3. PRIMARY TRIM MALFUNCTION.
    *Use the alternate trim system (alternate longitudinal trim levers)
    when trimming is required. Use the Autopilot as desired.*

♦ 4. PRIMARY LONGITUDINAL TRIM (ALL 3) C/B s
    (LEFT GENERATOR BUS PANEL) .... PULL
        ☆ CHECKLIST COMPLETE ☆

        | LONGITUDINAL TRIM INDICATOR CONTINUES TO MOVE | ⇐

3. STABILIZER TRIM (RED GUARDED SWITCH) ..... NORMAL

4. ALTERNATE TRIM MALFUNCTION.
    *Use the primary trim system (yoke thumb switch or longitudinal trim
    "suitcase" handles) to maintain the aircraft in trim.*

5. AUTOPILOT AND ALTERNATE LONGITUDINAL
    TRIM C/B s (D-9, D-10 & D-11) .... PULL

    CAUTION: THE AUTOPILOT MUST NOT BE USED
    DURING APPROACH OR DURING LARGE SPEED/
    CONFIGURATION CHANGES WITH THE ALTERNATE
    TRIM CIRCUIT BREAKERS PULLED. ENSURE THE
    AIRCRAFT IS IN TRIM BEFORE ENGAGING AUTOPILOT.

☆ RUNAWAY STABILIZER CHECKLIST COMPLETE ☆

MD-80 QUICK REF HANDBOOK   REV. - 11/15/99

1.17.2.3 Stabilizer Inoperative Checklist Procedures

The Alaska Airlines MD-80 QRH included the following checklist procedures to troubleshoot an inoperative stabilizer:

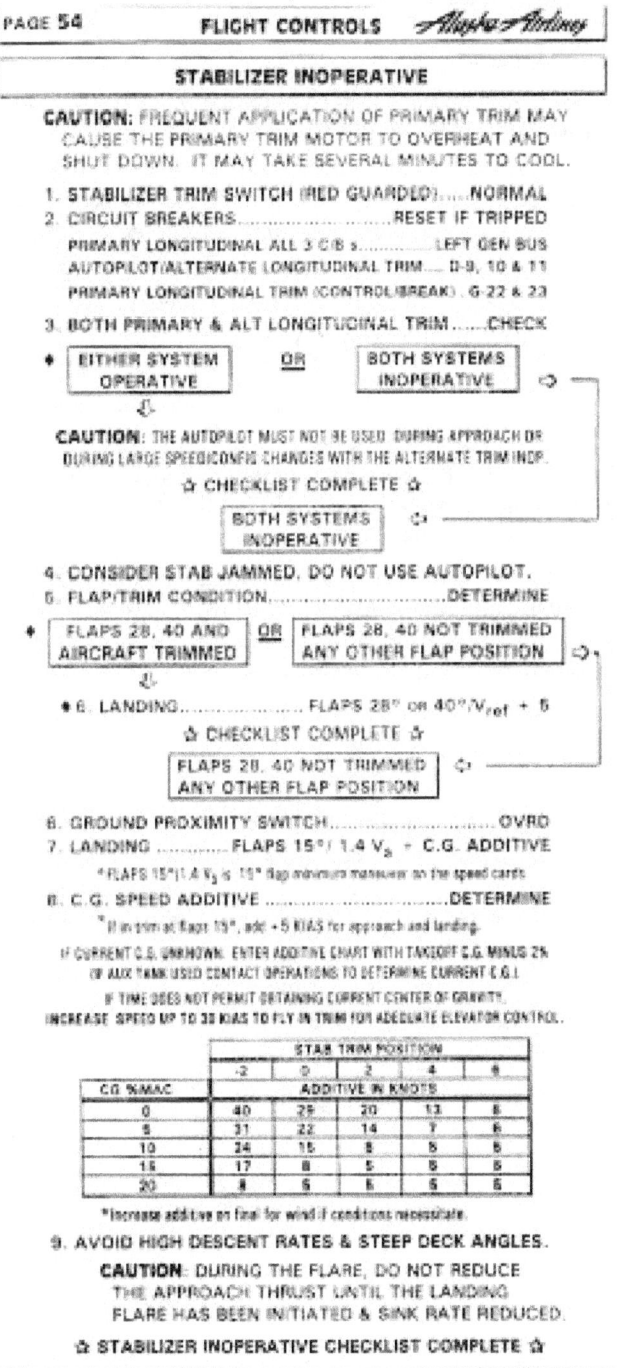

1.17.2.3.1 Postaccident Boeing Flight Operations Bulletin on Stabilizer Trim Inoperative/Malfunction Procedures

On February 10, 2000, Boeing issued a flight operations bulletin for DC-9, MD-80, MD-90, and 717 airplanes that addressed stabilizer trim inoperative/malfunction procedures. On the basis of information gathered in the flight 261 accident investigation to that date, the bulletin recommended the following:

> If a horizontal stabilizer trim system malfunction is encountered, complete the flight crew operating manual (FCOM) checklist(s). Do not attempt additional actions beyond that contained in the checklist(s). If completing the checklist procedures does not result in an operable trim system, consider landing at the nearest suitable airport. If an operable trim system is restored, the captain should consider proceeding to an airport where suitable maintenance is available, or to the original destination based on such factors as distance, weather, etc.

The Boeing bulletin added the following:

> The primary and alternate trim motors are each equipped with a thermal cut-off device which interrupts electrical current to the motor if that trim motor overheats. Repeated or continuous use of the trim motor may cause a thermal cutoff. After the motor cools, it will automatically restore trim function when the thermal cutoff resets. Because an overheat cutoff in one trim motor does not affect the functions of the other, if the alternate trim motor overheats, the primary trim system may be used to retrim the stabilizer; the reverse is also true. However, if the flight crew uses the primary trim system repeatedly to resist a runaway in the alternate trim system, the primary motor could overheat, and the crew may be left with a runaway alternate trim if the primary trim thermal cutoff occurs. This action could also cause both trim motors to overheat, and result in a temporarily inoperative stabilizer.

> Also, if a runaway trim motor overheats and stops, it could again runaway once it has cooled and the thermal cutoff resets. If the crew determines that the stabilizer is inoperative, and suspects that they may have a thermal cutoff, and if flight conditions permit, the captain may delay the diversion to an alternate airfield long enough to allow for a cooling period/thermal reset. After a reset, the crew should refer to the recommendations paragraph above.

The Boeing bulletin also advised flight crews that "excessive or prolonged testing of the trim system on the ground before departure may generate enough heat to produce a thermal cutoff during routine trimming shortly after takeoff."[160]

---

[160] After the flight 261 accident, there were reports of flight crews overtesting the trim systems, causing them to overheat.

1.17.3 FAA Oversight of Alaska Airlines

1.17.3.1 General

The Alaska Airlines operating certificate was issued in September 1946 and was managed by the FAA's Certificate Management Office (CMO) in Seattle, Washington. The CMO reported to the Seattle FSDO. The Seattle CMO staffing at the time of the flight 261 accident included a supervisor, five operations inspectors, and three principal inspectors: the PMI, the principal operations inspector, and the principal avionics inspector. The Seattle CMO staffing also included three geographic inspectors to assist with surveillance of Alaska Airlines: one from the Alaska Region, one from the Northwest Mountain Region, and one from the Western Pacific Region.

1.17.3.2 Preaccident FAA National Aviation Safety Inspection Program Inspection

From August 13 to August 23, 1995, a nine-member FAA National Aviation Safety Inspection Program (NASIP) inspection team conducted operations and airworthiness inspections at Alaska Airlines. During the inspection, the FAA team found "potential problems [with the airline's] systems for assuring compliance with FAR requirements" in the areas of "operations training, flight control, maintenance training programs, and contractual arrangements."

The final NASIP inspection report listed 16 noncompliance findings: 8 of the findings were related to operations, and the other 8 findings were related to airworthiness issues. Five of the operations findings were deficiencies in the flight attendant training manual, emergency drill training, and record-keeping procedures. Two operations findings were related to flight dispatcher and preflight briefings and guidance to flight crews related to weather issues. One of the airworthiness findings related to a maintenance action that was contrary to the certificate holder's manual and the FARs. The report stated that the mechanic tested the wrong system (the system tested was not the one reported to be inoperative) and that the mechanic was "not trained and qualified to perform maintenance on...[low visibility landing] maintenance program systems."

In addition, the report noted several deficiencies related to maintenance record-keeping and replacement part storage and found that the airline's GMM, which stated that repairs and overhauls could be conducted by another approved carrier, did "not contain a procedure, or a requirement, that will assure that the other carrier must meet the manufacturer's approved repair procedures."

1.17.3.3 FAA Air Transportation Oversight System

Alaska Airlines transitioned from the FAA's existing oversight system, the Program Tracking and Reporting System (PTRS), to the Air Transportation Oversight System (ATOS) in October 1998.[161] According to the FAA, ATOS was designed to focus

---

[161] The 10 largest U.S. air carriers were selected for the implementation of ATOS.

on "system failures" rather than "random events" and to improve the reporting and evaluation of system failures and trends for more aggressive intervention. Further, ATOS was designed to conduct inspections in 7 air carrier systems (aircraft configuration control, manuals, flight operations, personnel training and qualifications, route structure, flight rest and duty time, and technical administration), 14 air carrier subsystems, and 88 elements.

The ATOS program office manager testified that the transition to ATOS at some airlines, including Alaska Airlines, was difficult because of staff levels and training issues. He stated, "I think it would have been real prudent for us maybe to have overlaid some type of transition plan when we did implement ATOS, so that maybe we had a few of those comfortable things that worked well under our traditional surveillance still available when we implemented ATOS…We didn't really have that transition plan. We shut off the old one and started the new one." He also stated that, although some ATOS modules were "still at a rudimentary level of development…we're still far, far above where we were with just a compliance-based, event-based PTRS surveillance system. We have a lot better and a lot richer and a lot more robust data in that ATOS data repository right now than we've ever had under our traditional surveillance for these 10 carriers."

The former PMI for Alaska Airlines, who retired from the FAA in November 1999 after 8 1/2 years in that position, stated that the implementation of ATOS resulted "in a terrible transition." He also stated the following:

> The first thing that happened was the geographic support that we were accustomed to getting totally disappeared. We ended up with three other offices supporting us geographically; airworthiness, avionics and an operations inspector from each region that Alaska operated into. And between them coming up here and getting training and ATOS training and learning how to understand…all this stuff, they actually…[weren't] doing any surveillance. None of us were…we were too caught up doing ATOS things to actually go out and do any surveillance, do the system evaluations. They wouldn't let any geographic inspectors that weren't assigned to the carrier…look at the carrier. We used to get PTRS reports regularly from all over the field out there. We had about nine officers that were giving us geographic support. And after ATOS we ended up with about three officers, one in each region. And they were too busy doing all these complex…stuff. Nobody was out there looking at the carrier.

The Seattle CMO supervisor at the time of the ATOS implementation at Alaska Airlines stated that "the amount of surveillance that we have done since the introduction of ATOS has probably generally decreased. The [ATOS] concept is great. Translating that into some real life implementation is tougher than it sounds."

1.17.3.3.1 Seattle CMO Memorandum on Staff Shortages for Surveillance of Alaska Airlines

A November 12, 1999, memorandum, from the supervisor of the FAA's CMO in Seattle to the FAA's Director, Flight Standards Service, stated that staffing at the Seattle CMO "has reached a critical point" and added that "to accomplish our assigned activities, we require at least four additional airworthiness inspectors and one cabin safety inspector." The memorandum noted the following:

> [Alaska Airlines] one of the country's top ten air carriers, is undergoing a state of aggressive growth and expansion…We are not able to properly meet the workload demands. Alaska Airlines has expressed continued concern over our inability to serve it in a timely manner. Some program approvals have been delayed or accomplished in a rushed manner at the 'eleventh hour' and we anticipate this problem will intensify with time. Also, many enforcement investigations, particularly in the area of cabin safety, have been delayed as a result of resource shortages.

The memorandum stated that the Seattle CMO had implemented an ATOS comprehensive surveillance plan (CSP) and that the

> current CSP for airworthiness consists of 50 safety attribute inspections…and 176 element performance inspections…The CSP for operations consists of 50 [safety attribute inspections] and 103 [element performance inspections]. Effective execution of the ATOS inspections requires a significant expenditure of inspector resources. In order for the Seattle FSDO [CMO] to accommodate the significant volume of Alaska Airline's 'demand' work and effectively meet the objectives of the ATOS program, additional inspector staffing must be made available.

The memorandum concluded that, if the Seattle CMO "continues to operate with the existing limited number of airworthiness inspectors and without a cabin safety inspector, diminished surveillance is imminent and the risk of incidents or accidents at Alaska Airlines is heightened."

A May 3, 2002, memorandum from the FAA's Director, Flight Standards Service, in response to the Safety Board's request for information regarding staffing changes at the Seattle CMO since the flight 261 accident indicated that the authorized staffing level had increased from 15 to 37 and that the on-board staffing level had increased from 12 to 30 and would increase to 32 by May 19, 2002.[162] The memorandum indicated that the staffing included an operations research analyst and a cabin safety inspector.

1.17.3.4 FAA Postaccident Special Inspection of Alaska Airlines

As a result of the flight 261 accident, the FAA conducted a special inspection of Alaska Airlines from April 3 to April 19, 2000, to determine its compliance with the FARs. The inspection team was headed by a member of the FAA's System Process Audit staff from Washington, D.C., and team members included several FAA inspectors from the Seattle FSDO and CMO. According to the special inspection report, dated June 20,

---

[162] The Safety Board confirmed that the Seattle CMO on-board staffing level had increased to 32 by May 19, 2002, and determined that, as of September 2002, its on-board staffing level had further increased to 35.

2000,[163] the 15-member inspection team focused on Alaska Airlines maintenance and operations at its SEA and OAK facilities.

The special inspection report included the following findings:

Lack of Management Personnel:

- Director of Maintenance position has been vacant for nearly two years (since 6/12/98). Currently two individuals are sharing the position.
- Director of Operations position is currently vacant.
- The Director of Safety is also the Director of Quality Control and Training. This position also does not report directly to the highest level of management.

Maintenance Training:

- Alaska Airlines training] manual does not specify maintenance training curriculums or on-the-job training (OJT) procedures or objectives.
- The OJT program is informal, and is administered at the discretion of the appointed instructors. The program is not structured, there is no identification of subjects to be covered, and there are no criteria for successful completion provided.

Maintenance Program:

- General Maintenance Manual (GMM) does not reflect the procedures that the company is actually using to perform maintenance on its aircraft at the company's maintenance facilities.
- The GMM does not include 'how to' procedures regarding heavy check planning and/or production control.
- The GMM does not contain complete procedures for the issuance of an airworthiness release to an aircraft coming out of a heavy check.
- Two aircraft were released to service from a 'C' check without the completion of all necessary paperwork. This causes the inspection team to question the completion of all work on the aircraft before [it is] issu[ed] an airworthiness release and…used in revenue service.
- Spot checks of [Alaska Airlines'] shelf-life program for consumables revealed numerous discrepancies with the expiration dates that were exceeded or mislabeled.

---

[163] Details of the FAA special inspection report were officially released in a June 2, 2000, FAA press release.

- Shift turnover forms (MIG-48) are not being used consistently within the company as required by the GMM. Forms that are being used are incomplete and/or missing required signatures. There were also various other methods of shift turnover forms in use within the company that are not listed in the GMM.

- Numerous nonroutine work cards (MIG-4) were signed off using a company issued stamp, rather than an actual signature, as required by the GMM.

- Numerous MIG-4 forms had the box checked for 'partial work completed' but there were no entries on the back of the card for any partial work initiated or completed.

- Work cards being used for heavy checks are being modified or deleted without approval of engineering or quality control. Mechanics on the floor are making these changes on their own.

- There are no procedures in the GMM on how routine and nonroutine work cards will be distributed and controlled during a heavy check.

Continuous Analysis and Surveillance Program:

- [Alaska Airlines] manuals do not contain facsimiles of audit checklists to be used to administer the program.

- Data gathering is not continuous, but periodic. Audits are performed at 12 or 24-month intervals and an allowance for a six-month extension is available.

- Audit methods and techniques do not address compliance with regulatory safety standards. Rather, audit checklists are modeled after the CASE audit program, which is a generic program that is not specifically designed or tailored for [Alaska Airlines].[164]

The report also stated, "the area that showed the highest potential of systems breakdowns is in the Maintenance Program (15 findings). This is probably because Alaska Airlines maintenance personnel are not following the procedures that the company has in its manuals thereby increasing the probability for errors." The report added that "by not having a functional [continuing analysis and surveillance program], numerous other areas suffer from the lack of oversight and reform. This was evident within [Alaska Airlines]. It is controlled by the Quality Assurance department, which appears to be understaffed. Audits are not being completed in a timely manner and are incomplete, so problem areas are not being identified."

The special inspection report concluded the following:

- The procedures that are in place at [Alaska Airlines] are not being followed.

---

[164] The FAA's former director of Flight Standards Service stated at the Safety Board's public hearing that he was "staggered at the results of that inspection that showed that there were broad and systemic problems in the company's continuous analysis and surveillance program."

- The controls that are in place are clearly not effective, as measured by the number of findings that the team had during the inspection.

- The authority and responsibilities are not very well defined. This situation is aggravated [by] the fact that three positions [the director of maintenance, the director of operations, and the director of safety] are not filled. One of the positions, [director of maintenance], is being filled by two people, but the division of duties and responsibilities has not been made in the GMM; consequently, there is confusion as to who is responsible for what tasks.

- Control of the deferral system is missing. Items are being deferred without using the approved MEL [minimum equipment list]/CDL [configurations deviations list], resulting in items not being repaired for long periods of time.

- Quality Control and Quality Assurance Programs are ineffective. This is evident through things such as 'C' check packages that are missing signatures, open work cards, partial work completed, forms incomplete, etc.

The special inspection report's final assessment of Alaska Airlines' condition was that "most of the findings can be attributed to (a) the processes (both those in place and those that are missing) at the carrier and (b) ineffective quality control and quality assurance departments. Correcting these areas would eliminate the majority of the issues identified in this report."

The special inspection team also evaluated Alaska Airlines based on FAA criteria "for monitoring operators during periods of growth or major change." The inspection report stated that the airline planned to add additional aircraft, "which would result in a 20 percent increase over 3 years,"[165] and that the "company has a very high utilization rate, particularly with the MD-80 fleet, to accommodate its very aggressive flight schedule." The report noted that there was an "apparent shortage of [flight] instructors" at the airline, which "will become an even more prominent issue when aircraft are added to the fleet and more flight crews are needed, along with the increase in the summer schedule." The report further noted that because of a shortage of maintenance personnel at OAK, "contractors will be used more for accomplishing their 'C' checks" and that "Quality Assurance…and Quality Control…are not adequately staffed at contract maintenance facilities." The report suggested that the key elements of Alaska Airlines' system, including the maintenance deferrals, quality assurance and control, internal evaluation, maintenance control, and GMM "should be monitored to determine if the carrier's system is able to support additional aircraft."

1.17.3.4.1 FAA Proposed Suspension of Alaska Airlines' Heavy Maintenance Authority

In its June 2, 2000, press release regarding the special inspection findings, the FAA proposed the suspension of Alaska Airlines' heavy maintenance authority. The FAA press

---

[165] These criteria are found in handbook bulletins HBAW 98-21 and HBAT 98-36. The inspection team's report noted that, according to these handbook bulletins, "a growth rate of 10 percent or more is high."

release stated, "approximately six to seven aircraft, of the airline's fleet of 89 aircraft, are in heavy maintenance in any given month." The FAA stated that Alaska Airlines was working with the FAA "to correct the deficiencies outlined in the inspection." On June 29, 2000, the FAA accepted an Airworthiness and Operations Action Plan submitted by Alaska Airlines, and its authority to conduct heavy maintenance was not suspended.

1.17.3.4.2 Alaska Airlines Airworthiness and Operations Action Plan

Alaska Airlines' June 29, 2000, Airworthiness and Operations Action Plan stated that the airline had "instituted both interim and long-term measures that will ensure that its maintenance operations meet or exceed all Federal Aviation Regulations and that all aircraft released from heavy maintenance are safe and airworthy with all maintenance properly documented." The action plan stated that the airline had "initiated a top-to-bottom review of all operations" and was developing a continuing analysis and surveillance program "that will not only detect and correct any deficiencies in Alaska's maintenance program, but will constantly improve the program through a process of data collection, analysis and corresponding changes."[166]

The action plan stated that the Alaska Airlines maintenance training program had been reorganized to develop a "training profile for all maintenance personnel by area" and that OJT would be "standardized and formalized." The action plan stated that Alaska Airlines would take the following actions regarding its maintenance program:

> audit 100 percent of the maintenance documents produced during a heavy maintenance check, for every airplane, at every location, whether accomplished in house or by an approved vendor. We also conduct random audits on every airplane during each heavy check. We analyze the findings to validate the new heavy maintenance procedures. If the analysis indicates a problem with a new procedure, it is corrected immediately…heavy maintenance procedures contained in our GMM have been completely revised utilizing a meticulous process implemented by Alaska Airlines personnel in conjunction with highly-qualified outside consultants under the supervision of the FAA.

Further, the action plan stated that Alaska Airlines had "created an executive level safety position, vice president of safety, that reports directly to our CEO." In addition, the plan stated that Alaska Airlines had created more than 130 new positions in its maintenance and engineering division, 28 new positions in its flight operations division, and 11 new positions in its safety division.

The action plan also stated that new procedures and policies had been implemented to improve monitoring of contract maintenance vendors and "to ensure that all heavy maintenance performed at a vendor meets or exceeds requirements of applicable

---

[166] A Department of Transportation (DOT) Office of the Inspector General report on FAA oversight of continuing analysis and surveillance programs, dated December 12, 2001, concluded that, if the continuing analysis and surveillance program "had been operating effectively, Alaska Airlines' own internal monitoring process should have identified the deficiencies in its maintenance program." The FAA acknowledged the findings in the DOT report. For more information about the DOT report, see section 1.17.3.4.4.

FARs and the Alaska maintenance program." The plan added that a policies and procedures manual was developed for heavy maintenance vendors to "ensure consistency…and consideration of industry best practices. This manual provides the basis for vendor representative accountability, consistency of policies and training of new vendor representatives."

Finally, the action plan stated that Alaska Airlines had developed an internal audit program "to provide oversight of all operating departments within Alaska Airlines…[and] to develop an effective process to validate major changes to process or procedures at Alaska Airlines." The plan stated that this internal audit would be "mandatory for flight operations, in-flight services, maintenance and engineering and airport operations."

1.17.3.4.3 FAA Followup Evaluation

During the week of September 18, 2000, the FAA conducted a focused followup evaluation of Alaska Airlines' SEA and LAX maintenance facilities and a contract maintenance facility in Everett, Washington.

According to a September 22, 2000, FAA memorandum, the purpose of the evaluation was to "assist the [FAA's] Alaska Airlines Certificate Management Office (CMO) in completing an assessment of the carrier's revised heavy maintenance process, nonroutine work cards (MIG-4), vendor maintenance process, quality assurance program, and continuing analysis and surveillance." The memorandum added that "during the second day of this evaluation it was determined that the new programs were not completely implemented and the scope of the evaluation would be limited to a 'progress check.'"

The September 22, 2000, memorandum listed procedural discrepancies and several recurring findings, including the following:

> Alaska Airlines does record and transmit information and instructions to their counterparts for shift turnovers as required by the GMM. Two Alaska Airlines forms titled, Priority Items and Engineering Report, are being utilized during shift turnovers. Alaska Airlines has neither identified the aforementioned forms [n]or implemented procedures in their [GMM] for the use of the forms.

> Not all of the quality assurance (QA) auditors have been trained on the new [continuing analysis and surveillance] procedures, and the company personnel that were explaining the new [continuing analysis and surveillance] procedures to us [the evaluation team] did not appear to understand their own process of qualifying and authorizing QA auditors.

> There is no formal course or syllabus for [continuing analysis and surveillance] auditor training. The training program consists of only informal instruction accomplished during OJT. There are no checklists or guides being used to standardize the OJT. There are no controls or testing requirements in place to determine how much the students were comprehending and retaining.

The September 22, 2000, memorandum also noted that Alaska Airlines appeared not to have "implemented procedures for documenting the qualifications and authorizations of QA auditors" and that "there was a lack of standardization between facilities, both internal and external, on following the instructions for reference material attached to the MIG-4." The memorandum concluded that "these [finding] areas should be re-visited after the manuals have been re-written, the programs have been fully implemented, and all personnel have been trained. This is projected to be in January 2001."

A September 25, 2000, FAA memorandum summarized an inspection team evaluation of Alaska Airlines' maintenance reliability program document, which was the basis of the airline's reliability analysis program.[167] The memorandum listed irregularities concerning the document, which included that the document did not

- define all of the aircraft components controlled by the program,
- identify the individual responsible for analyzing trend-related information,
- address the individuals responsible for conducting further analysis for corrective action of program deficiencies or negative trends as identified by the program's analytical process,
- contain a specific method to follow so that data are routed to the proper organizational element for review,
- contain a method for ensuring that changes in operating flight hours or cycles as appropriate to the trend level of reliability experienced occurs,
- contain a procedure to assign the appropriate personnel to find the cause of all identified areas that exceed performance standards,
- contain a method for establishing extension limits, and
- contain a method for determining the analysis of nonroutine maintenance work forms as a significant analytical factor in...maintenance program effectiveness.

The September 25, 2000, memorandum concluded the following:

Of concern is that the maintenance reliability program [document] adequately depicts an accurate portrait of the maintenance program's total effectiveness in a single source document. And further, that changes to the maintenance as identified through the reliability program analysis, are reviewed, controlled, and FAA CMO-approved as warranted. The validity of a maintenance reliability program, as reflected in Operations Specifications, does not imply carte blanche to an operator for maintenance program changes without sufficient analytical data presented as justification. The analytical data should accompany any notice or request for a maintenance program change. Sufficient FAA guidance in publications exists in support of this assertion.

---

[167] For a description of Alaska Airlines' reliability analysis program, see section 1.6.3.1.3.

In July 2001, an FAA 12-member panel again reviewed Alaska Airlines' progress in implementing its action plan.[168] According to an FAA briefing document on the review, the review consisted of presentations by Alaska Airlines representatives, a visit by some of the panel members to Alaska Airlines' SEA maintenance facility, and discussions about Alaska Airlines' progress. According to the briefing document, Alaska Airlines "convincingly demonstrated [its] completion of all initiatives and commitments set forth in their action plan." The briefing document indicated that the panel members who visited the SEA maintenance facility found that, "with one exception, Alaska Airlines technicians [had] an acceptable working knowledge of the automated General Procedures Manual." Further, the briefing document stated that the panel had "reached a consensus that Alaska Airlines met or exceeded all commitments set forth in their action plan...[and that the airline had demonstrated] strong and visible leadership from the top executives…for safety, compliance and a strong and healthy safety culture."

The FAA briefing document stated that Alaska Airlines had added 341 maintenance positions, enhanced its internal evaluation program, created a Board of Directors' Safety Committee "that interfaces with the director of safety," improved General Procedures Manual written procedures, and revised "flight operations procedures and corresponding training that address critical operational control issues and the interface between maintenance and flight operations with emphasis on abnormal and emergency procedures." The briefing document concluded that Alaska Airlines had "demonstrated…that previously identified systemic deficiencies have been corrected and verified as corrected" and that the FAA had verified "significant changes and improvements at Alaska." The briefing document was signed by the manager of the FAA's Northwest Mountain Region Flight Standards Division.

1.17.3.4.4 DOT Office of the Inspector General Report on FAA Oversight of Continuing Analysis and Surveillance Programs

In its December 12, 2001, report on FAA oversight of continuing analysis and surveillance programs, the DOT Office of the Inspector General stated that the findings of the FAA's postaccident inspection of Alaska Airlines "raised questions as to why the FAA's routine surveillance had not identified the deficiencies in Alaska Airlines' [continuing analysis and surveillance program] and ensured that they were corrected." The DOT report stated that the FAA "needs to place greater emphasis on [continuing analysis and surveillance program] oversight" and must "ensure [that continuing analysis and surveillance program] deficiencies identified through its oversight inspections are corrected."

---

[168] Four of the panel members were participants in the April 2000 special inspection.

The DOT report added the following:

> The lack of effective oversight of air carriers' [continuing analysis and surveillance programs] perpetuates a system where [the] FAA relies on its own inspections to ensure carriers maintain their aircraft in an airworthy condition. This system is ineffective because the FAA does not have sufficient resources to physically inspect every aircraft. While it is clearly the responsibility of air carriers to ensure day-to-day safe operation and maintenance of their aircraft, [the] FAA must be more proactive in identifying deficiencies in air carriers' [continuing analysis and surveillance programs].

The DOT report also noted that to properly evaluate a maintenance program, an airline should have the following:

> procedures in its [continuing analysis and surveillance program] to assess aircraft mechanical performance. For example, the carrier should review data such as engine removal rates and pilot reports of mechanical disruptions to identify negative trends or premature failures. Mechanical monitoring programs can help carriers maintain reliable aircraft operating rates by identifying [the] cause of maintenance-related delays and cancellations. More importantly, a properly designed and utilized [continuing analysis and surveillance program] establishes a culture of safety with an airline's operations.

The DOT report concluded that FAA inspectors "need to better document their [continuing analysis and surveillance program] inspections" and "better train its inspectors to evaluate carriers' [continuing analysis and surveillance programs] for systemic weaknesses."

1.17.4 Other Safety Evaluations of Alaska Airlines

1.17.4.1 DoD Capability Survey of Alaska Airlines

In September 1998, the DoD's Air Carrier Survey and Analysis Office conducted a biennial capability survey of Alaska Airlines' operations.[169] The DoD survey report stated the following:

> The commendable practices we observed during the survey were Alaska Airlines' overall training program to include the excellent crew coordination exhibited during DoD cockpit observations. Operations management conducts ongoing internal audits that include effective cross-divisional audits. Alaska's safety program also exceeds standards with a highly motivated staff and exceptional projects like their flight operational quality assurance…program.

Regarding operations management, the DoD survey report stated that Alaska Airlines had a "well-defined organizational structure, with staffing levels commensurate

---

[169] DoD capability surveys of air carriers are part of a routine approval process for participation in the DoD's air transportation program.

for their scope of operations" and "very experienced personnel in key positions with minimal management turnover." Regarding maintenance management, the survey report stated that Alaska Airlines management provided "strong oversight of maintenance" and that "company structure and lines of authority [were] satisfactorily defined." The report added that "excellent communications exist[ed] with both operations and mechanics."

The DoD survey report stated that Alaska Airlines had a "strong safety program that is responsive to employee inputs" and that "interaction between safety, functional managers, and unions occurs frequently through formal meetings, written communications, and informal discussion." The survey report further stated that Alaska Airlines "[continuing analysis and surveillance] program provides strong oversight of all maintenance activities" and that a formal internal audit program provided "excellent supervision of the company's internal maintenance functions." The report noted that "during summer 1998, [the] carrier experienced difficulty accomplishing some heavy checks on time due to high summer operations tempo and fluctuating heavy check flow." The report concluded that the Alaska Airlines' operations and maintenance management and operations safety program and training "exceed [DoD] standards."

1.17.4.2 Criminal Investigation of Alaska Airlines

In October 1998, an Alaska Airlines senior lead mechanic alleged to the FAA that maintenance records had been falsified at Alaska Airlines' OAK maintenance facility. His statements triggered a 3-year criminal investigation into the allegations by a Federal grand jury. In December 2001, Federal prosecutors stated that no charges would be filed against Alaska Airlines as a result of the investigation.

1.17.4.3 Postaccident Independent Safety Assessment of Alaska Airlines

After the accident, Alaska Airlines hired a consulting firm to conduct an independent safety assessment of the airline. The firm assembled a 13-member team of subject matter experts in flight operations, hazardous materials handling, security, and maintenance and engineering. These areas included organizational structure and training, flight dispatch, crew scheduling, station and ramp operations, cabin safety, maintenance quality assurance programs, maintenance records systems, mechanical reliability reporting, and maintenance contracting. The assessment team's inspections of Alaska Airlines were conducted from April 10 to May 19, 2000.

The final safety assessment report, dated June 19, 2000, stated that the purpose of the assessment was to examine "the prevailing operational situation from a safety perspective (i.e., the 'safety culture') and to identify areas where Alaska Airlines' overall operational safety and efficiencies could be enhanced or improved through strengthening or modifying existing policies, procedures or practices, or through new initiatives." The report added that the assessment focused on "examination of safety issues from the standpoint of organizational functionality and management oversight and the evaluation of Alaska Airlines' internal policies, procedures and practices."

The safety assessment team made more than 170 recommendations that it stated were intended to "guide both strategic planning and tactical action by Alaska Airlines as they move to strengthen the foundation of an effective safety culture that anticipates the potential for problems and acts swiftly and effectively to mitigate such threats." The recommendations suggested, in part, that Alaska Airlines develop clearer lines of managerial authority and responsibilities; increase staffing; enhance standardization of operating and training manuals; revise its GMM to clarify and update policies, responsibilities, and procedures; increase maintenance training; and create a new corporate safety department that reports directly to the highest levels of the airline. The report also recommended that Alaska Airlines appoint a vice president of corporate safety to manage the department.

1.18 Additional Information

1.18.1 Postaccident Airworthiness Directives

On February 11, 2000, as a result of information gathered immediately after the flight 261 accident, the FAA issued AD 2000-03-51 to all operators of DC-9, MD-90-30, MD-88, and 717-200 series airplanes. AD 2000-03-51 stated the following:

> [after the Alaska Airlines crash, the] FAA…received a report from an operator that indicated two instances of metallic shavings in the vicinity of the jackscrew assembly and gimbal [acme] nut of the horizontal stabilizer. Metallic shavings in the vicinity of the horizontal stabilizer indicate excessive wear of the jackscrew assembly. Such excessive wear, if not corrected, could result in possible loss of pitch trim capability, which could result in loss of vertical control of the airplane.

AD 2000-03-51 required operators to "perform a general visual inspection of the lubricating grease on the jackscrew assembly and the area directly below the jackscrew and surrounding areas for the presence of metal shavings and flakes" and to replace the assembly before further flight if shavings or flakes were found.[170] The AD also required inspection and lubrication of the jackscrew assembly, "prior to the accumulation of 650 hours total time-in-service…or within 72 hours" after receipt of the AD and were to be repeated "at intervals not to exceed 650 flight hours."

In addition, AD 2000-03-51 required operators to "perform a general visual inspection of the jackscrew assembly to detect the presence of corrosion, pitting or distress" and to replace the assembly if these conditions were found. Further, the AD required an inspection of the lubrication of the jackscrew assembly, an inspection of the acme screw and nut upper and lower mechanical stops, testing of the horizontal stabilizer shutoff control, and the performance of end play checks "within 2,000 flight hours since

---

[170] On February 11, 2000, Boeing issued ASBs for DC-9, MD-90-30, MD-88, and 717-200 series airplanes that described the "procedures for inspecting the general condition of the jackscrew assembly and the area around the jackscrew assembly to detect the presence of metal shavings and flakes."

the last acme screw and nut wear check" and, thereafter, to be repeated at intervals not to exceed 2,000 flight hours.

On July 28, 2000, the FAA issued AD 2000-15-15, which superseded AD 2000-03-51. AD 2000-15-15, which had an effective date of August 23, 2000, included the same requirements as AD 2000-03-15, but AD 2000-15-15 also required operators to look for the presence of metallic particles in the jackscrew assembly lubrication. The AD stated the following:

> Since the issuance of AD 2000-03-51, the FAA has received numerous reports of incidents in which metallic particles (including slivers and dust, as well as shavings and flakes) were found [e]mbedded within the grease on the threaded portion of the jackscrew assembly of the horizontal stabilizer actuator and on the area directly below the jackscrew assembly. Findings by the manufacturer indicate that such metallic particles can be identified as a non-magnetic metallic substance which is golden in color...in consideration of new findings by the manufacturer regarding the types of material found in the jackscrew assembly of the horizontal stabilizer since issuance of AD 2000-03-51, the FAA has determined that the required inspections should be expanded to include metallic particles such as slivers and dust, as well as the metal shavings and flakes identified in AD 2000-03-51.

AD 2000-15-15 concluded that "the FAA has determined that it is necessary for operators to report the results of the end play checks" performed at the 2,000-flight-hour intervals prescribed in AD 2000-03-51 "to provide information regarding the wear rates of the jackscrew assembly." The AD stated that the FAA would use this reported end play data "to confirm that the repetitive intervals of 650 flight hours [for the general visual inspection of the jackscrew assembly and lubrication]...and the repetitive intervals of 2,000 flight hours [for the end play checks]...are appropriate compliance times for accomplishment of the end play check and are adequate for ensuring the safety of the fleet."

1.18.1.1 Safety Board and FAA Correspondence Regarding the 2,000-Flight-Hour End Play Check Interval Specified in ADs 2000-03-51 and 2000-15-15

In a May 14, 2001, letter to the FAA, the Safety Board indicated that the results of a February 17, 2000, end play check on a Hawaiian Airlines DC-9,[171] which yielded an end play measurement of 0.043 inch after having accumulated only about 2,500 flight hours, called into question the adequacy of the 2,000-flight-hour end play check interval required by AD 2000-15-15. The Board stated that, because of the wear observed on the Hawaiian Airlines jackscrew assembly and the potential that other assemblies with higher-than-normal wear rates are in service, it was concerned that the 2,000-flight-hour end play check interval might not be adequate to ensure the safety of the fleet affected by AD 2000-15-15. The Board requested any additional information regarding the background data and analysis that the FAA used to establish the 2,000-flight-hour end

---

[171] For more information about the Hawaiian Airlines DC-9 jackscrew assembly, see section 1.16.6.1.

play check interval identified in ADs 2000-03-51 and 2000-15-15 and any final or terminating actions it had identified in this case.

In a June 26, 2001, letter, the FAA responded that Boeing initially recommended the 2,000-flight-hour end play check interval in ASB DC-9-27A362. The FAA stated that, while evaluating this recommendation, it had numerous discussions with Boeing regarding the acceptability of the 2,000-flight-hour interval. The FAA added the following:

> These discussions focused on the robust design of the acme nut, structural test data, structural analysis, existing wear rate data, and the service history of the fleet. The original Boeing wear rate data showed a wear rate of 0.001 inch per 1,000 flight hours when the jackscrew was lubricated at 400-hour intervals. The Boeing data was supplemented with recent wear rate information provided to Boeing by one of the largest MD-80 operators. These new data showed a wear rate which closely correlated with the existing Boeing wear rate but was obtained from aircraft with jackscrews that were being lubricated at approximately 940-flight-hour intervals…analysis showed that a jackscrew worn beyond 0.080-inch end play could still safely carry limit load. Based on FAA review of the analytical data and on the discussions with Boeing, we concluded that Boeing's recommendation for endplay checks at 2,000-flight-hour intervals supported by lubrication and inspections at 650-flight-hour intervals was sufficient to produce predictable wear rates and detect worn jackscrew assemblies well before their structural capability would be compromised.

In response to the Safety Board's concern that the 2,000-flight-hour end play check interval would be inadequate to identify a jackscrew assembly that is wearing at a higher-than-normal rate, the FAA stated, in part, the following:

> The FAA is also concerned about jackscrews found with excessive wear rates because these are seen as indicators not only of possible improper lubrication or contaminated grease but also of a potential design problem. With one exception…the current endplay check intervals have shown to be adequate to detect jackscrews with accelerated wear rates….If [the Hawaiian Airlines] jackscrew were to have remained in service…and if it continued to wear at the same rate, the lubrication and inspections requirements of AD 2000-15-15 would have provided at least two opportunities for maintenance personnel to detect excessive wear before the next required endplay check. This is because jackscrews wearing at an excessive rate will produce an excessive amount of debris. The AD requires an inspection for debris at each lubrication interval and, if found, a requirement to perform an endplay check. Were the excessive wear not detected before the next endplay check, the jackscrew assembly with the amount of endplay identified in your letter…would retain sufficient structural capability to safely support limit load.

In response to the Safety Board's query about whether any terminating actions had been identified on this issue, the FAA stated, in part, the following:

The FAA has had discussions with Boeing regarding the effectiveness of the existing inspections and the appropriateness of design changes to the jackscrew assembly. To date, these discussions have yielded no firm conclusions that a design change is either appropriate or warranted.

1.18.2 Safety Board Statistical End Play Data Study

The Safety Board statistically examined the end play data collected as a result of ADs 2000-03-51 and 2000-15-15. The Board received a total of 3,174 end play measurements from 1,493 airplanes at 44 operators.

Because end play was measured at repeated intervals for individual jackscrew assemblies, it was possible to estimate whether the end play measurement was reliable.[172] According to Boeing estimates, very little change in end play was expected to occur over the 2,000-flight-hour interval between checks because end play was expected to increase at a rate of about 0.001 inch per 1,000 flight hours. Because this expected change was small and unidirectional, the Safety Board used the test-retest method to estimate measurement reliability.[173]

End play measurements reported on three consecutive occasions[174] for the same airplanes were compared using a correlation analysis that measures the strength of the relationship between two sets of numbers and produces a value known as the correlation coefficient (represented in notation as "r"). The correlation coefficient ranges between ±1.0.[175] The correlation coefficient for the first and second measurements was +0.553 and +0.416 for the second and third measurements. The coefficients of determination, (represented in notation as "R^2"), representing shared, or common, variance, were 0.306 for the first and second measurements and 0.173 for the second and third measurements, indicating low measurement reliability.[176]

In addition to analyzing on-wing end play measurements collected as a result of ADs 2000-03-51 and 2000-15-15, a second portion of the end play data study focused on

---

[172] Reliability refers to the ability of a measurement tool to yield reproducible results.

[173] Measurement reliability is often assessed using a test-retest method in which two consecutive measurements are recorded and, if the entity being measured has not changed (or has changed in a consistent manner) during the measurement interval, subsequent correlation of the two measurements should reveal a strong relationship. The absence of such a correlation suggests that other variables, such as measurement error, may have caused variability in the observed measurement.

[174] After screening the data for errors (missing data, duplicate cases, multiple measurement conventions, incorrect flight hours, and incorrect fuselage numbers), the resulting set contained 1,388 cases at the first measurement; 852 cases at the second measurement; and 482 cases at the third measurement.

[175] A coefficient of +1.0, known as a perfect positive correlation, means that in each case changes in one measurement resulted in an identical change in the other measurement. A coefficient of -1.0, known as a perfect negative correlation, means that changes in one measurement resulted in an identical change in the other measurement but that the change was in the opposite direction. A coefficient of 0 means that no relationship existed between the two measurements and that changes in one measurement had no effect on the second measurement.

[176] Two alternative test-retest analyses that accounted for varying time intervals between checks for individual aircraft were performed and yielded similar results.

an examination of a small subset of cases in which both on-wing and bench-check end play measurement data were available. Boeing collected and reported data on jackscrew assemblies that were returned to Integrated Aerospace during 2000. Integrated Aerospace reported that 157 jackscrew assemblies were returned for overhaul during the 2000 calendar year, a marked increase over those received in previous years,[177] and likely affected by the AD requiring end play checks for the entire fleet of DC-9, MD-90-30, MD-88, and 717-200 series airplanes. Of the 157 jackscrew assemblies, 49 were removed because of end play measurements that exceeded operators' criteria for removal. Other reasons for removal included the presence of metal flakes, excessive free play, or damage to the assembly.

Jackscrew assemblies returned to Integrated Aerospace are cleaned and then bench end play checks are conducted. Because some of the conditions that can create variation in the end play measurement are controlled during bench checks, they are expected to yield more accurate end play measurements than on-wing checks conducted by maintenance personnel. Of the 157 jackscrew assemblies removed in 2000, bench end play checks for 142 of the assemblies resulted in end play measurements that fell within in-service end play tolerances (between 0.003 and 0.040 inch). Twelve assemblies had bench-check end play measurements greater than 0.040 inch. No bench-check end play measurements were reported for three assemblies. Boeing contacted operators and obtained a matched sample (both on-wing and bench-check end play measurements) for 64 jackscrew assemblies.

After correcting documentation errors, the Safety Board calculated the difference in end play measurements between the on-wing and the bench-check settings. In 45 of the 64 cases (70.3 percent), the on-wing end play measurements were higher than the bench-check measurements; in 6 of the 64 cases (9.4 percent), the on-wing and bench-check end play measurements were equal; and in 13 of the 64 cases (20.3 percent), the on-wing end play measurements were lower than the bench-check measurements.

After removing one outlier,[178] the validity of the on-wing end play measurement was approximated using a correlation analysis between the on-wing and bench-check end play measurements.[179] The analysis determined that the resulting correlation coefficient was +0.442 (R^2=0.195), suggesting low measurement validity. The Safety Board end play data study report, dated March 18, 2002, concluded, "in the absence of additional information such as the rate of acme nut wear and the thread thickness at which failure may occur, the observed level of measurement error raises doubts about the utility of the existing end play measurement procedure."

---

[177] According to Integrated Aerospace records, 59 jackscrew assemblies were returned in 1999, 54 in 1998, 59 in 1997, and 42 in 1996.

[178] An outlier is an extreme measurement that is not representative of the rest of the data set.

[179] This method assumed that the bench-check end play measurement was a criterion, or a standard, that represented the true end play in the jackscrew assembly. If the on-wing end play measurement is valid (representing the actual state of end play), then the correlation coefficient representing the relationship between the on-wing and bench-check end play measurements should be close to +1.0.

1.18.3 Previous Safety Recommendations Resulting from the Alaska Airlines Flight 261 Investigation

On October 1, 2001, the Safety Board issued Safety Recommendations A-01-41 through -48. In the safety recommendation letter, the Board expressed concern that lubrication procedures for the acme screw and nut were not sufficient to ensure thorough lubrications and that end play check procedures were not adequate to ensure accurate measurements of acme screw and nut wear. In addition, the Board expressed concern that use of an inappropriate grease type or a mixture of incompatible grease types could cause adverse effects and that technical information and practical experience about aviation lubrication was not being shared by the aviation industry.

Safety Recommendation A-01-41 asked the FAA to do the following:

Require the Boeing Commercial Airplane Group to revise the lubrication procedure for the horizontal stabilizer trim system of Douglas DC-9, McDonnell Douglas MD-80/90, and Boeing 717 series airplanes to minimize the probability of inadequate lubrication.

In a December 12, 2001, response letter, the FAA stated that it agreed with the intent of Safety Recommendation A-01-41 and that it was "working with the Boeing Commercial Airplane Group to rewrite the lubrication procedures to the optimal standard." The FAA also stated that it was "evaluating the necessity of an additional task to perform a detailed inspection of the lubricant at a 'C' check interval to check for indications of metal shavings."

On June 14, 2002, the Safety Board classified Safety Recommendation A-01-41 "Open—Acceptable Response," pending the revisions to the lubrication procedure for the horizontal stabilizer trim system of DC-9, MD-80/90, and 717 series airplanes. On July 25, 2002, Boeing representatives provided the Board with a briefing on new lubrication procedures currently being developed and tested.

Safety Recommendation A-01-42 asked the FAA to do the following:

Require the Boeing Commercial Airplane Group to revise the end play check procedure for the horizontal stabilizer trim system of Douglas DC-9, McDonnell Douglas MD-80/90, and Boeing 717 series airplanes to minimize the probability of measurement error and conduct a study to empirically validate the revised procedure against an appropriate physical standard of actual acme screw and acme nut wear. This study should also establish that the procedure produces a measurement that is reliable when conducted on-wing.

In its December 2001 response letter, the FAA stated that it agreed with the intent of Safety Recommendation A-01-42 and that it had asked Boeing to "revise the end play check procedure…to minimize the probability of measurement error." The FAA further stated that it had asked Boeing "to conduct a study to validate the revised procedure empirically against an appropriate physical standard of actual acme screw and acme nut wear. The study will also establish procedures that produce a measurement that is reliable when conducted on-wing."

On June 14, 2002, the Safety Board classified Safety Recommendation A-01-42 "Open—Acceptable Response," pending the revisions to the end play check procedure and the validation of the revised procedure. On July 25, 2002, Boeing representatives provided the Board with a briefing on revised end play check procedures currently being developed and tested.

Safety Recommendation A-01-43 asked the FAA to do the following:

> Require maintenance personnel who lubricate the horizontal stabilizer trim system of Douglas DC-9, McDonnell Douglas MD-80/90, and Boeing 717 series airplanes to undergo specialized training for this task.

Safety Recommendation A-01-44 asked the FAA to do the following:

> Require maintenance personnel who inspect the horizontal stabilizer trim system of Douglas DC-9, McDonnell Douglas MD-80/90, and Boeing 717 series airplanes to undergo specialized training for this task. This training should include familiarization with the selection, inspection, and proper use of the tooling to perform the end play check.

In its December 2001 response letter, the FAA stated that it believed that "current regulatory requirements of 14 CFR 121.375 adequately address maintenance training programs." The FAA added that the agency was "revising [AC] 120-16C, *Continuous Airworthiness Maintenance Programs*, to expand the intent of the requirement for maintenance training programs." The FAA stated that the AC revisions would address basic training, initial training, recurrent training, specialized training, vendor training, and competence-based training. In addition, the FAA asserted that "current regulatory requirements of 14 CFR 65.81 and 14 CFR 65.103 adequately address the inspection requirements for maintenance personnel and repairmen." The FAA concluded that "current regulatory requirements in place to require maintenance training programs and inspection requirements for maintenance personnel and repairmen address the full intent of these safety recommendations."

In its June 14, 2002, response letter to the FAA, the Safety Board stated that it "continues to believe that current FAA regulations do not result in maintenance personnel being properly trained to perform lubrication and end play inspection of the acme screw and nut assembly" and urged the FAA to reconsider its position. The Board classified Safety Recommendations A-01-43 and -44 "Open—Unacceptable Response," pending accomplishment of the recommended actions.

Safety Recommendation A-01-45 asked the FAA to do the following:

> Before the implementation of any proposed changes in allowable lubrication applications for critical aircraft systems, require operators to supply to the FAA technical data (including performance information and test results) demonstrating that the proposed changes will not present any potential hazards and obtain approval of the proposed changes from the principal maintenance inspector and concurrence from the FAA applicable [Aircraft Certification Office].

Safety Recommendation A-01-46 asked the FAA to do the following:

> Issue guidance to principal maintenance inspectors to notify all operators about the potential hazards of using inappropriate grease types and mixing incompatible grease types.

In its December 2001 response letter, the FAA stated that it was "issuing a flight standards information bulletin to address these safety recommendations." The FAA added that the bulletin would do the following:

> provide inspectors with guidance on processing an operator's proposed substitution of a lubricant or a lubrication application method change. The bulletin will direct aviation safety inspectors to request that the operator supply technical data on the substitute lubricant or [lubrication application method change] that affects critical aircraft systems and include performance information, along with test results, demonstrating that the proposed change will not present any potential hazard. The bulletin will also direct aviation safety inspectors to have the analysis test data reviewed by the appropriate FAA Aircraft Certification Office to determine if the substitute lubricant meets the required specifications.

In addition, the FAA stated that it was "incorporating a revision in Order 8300.10, Airworthiness Inspector's Handbook, to state that if there is any doubt as to the soundness of the request, the aviation safety inspector should coordinate the request with the appropriate Aircraft Certification Office." In an April 2, 2002, letter, the FAA further responded to Safety Recommendations A-01-45 and -46, stating that, on February 15, 2002, it had issued Flight Standards Information Bulletin for Airworthiness (FSAW) 02-02, "The Potential Adverse Effects of Grease Substitution" and revised Order 8300.10 per its previous letter.

On July 29, 2002, the Safety Board classified Safety Recommendations A-01-45 and -46 "Closed—Acceptable Action" based on the issuance of FSAW 02-02 and the revisions to Order 8300.10.

Safety Recommendation A-01-47 asked the FAA to do the following:

> Survey all operators to identify any lubrication practices that deviate from those specified in the manufacturer's airplane maintenance manual, determine whether any of those deviations involve the current use of inappropriate grease types or incompatible grease mixtures on critical aircraft systems and, if so, eliminate the use of any such inappropriate grease types or incompatible mixtures.

Safety Recommendation A-01-48 asked the FAA to do the following:

> Within the next 120 days, convene an industrywide forum to disseminate information about and discuss issues pertaining to the lubrication of aircraft components, including the qualification, selection, application methods, performance, inspection, testing, and incompatibility of grease types used on aircraft components.

In its December 2001 response letter, the FAA stated that it agreed with the intent of Safety Recommendations A-01-47 and -48 and had scheduled an FAA/industry forum that would include "representatives from domestic and international airlines [and] will address the issues raised in these safety recommendations." The FAA/industry forum was held in January 2002.

In its June 14, 2002, response letter to the FAA, the Safety Board requested that the FAA provide copies of any proceedings or recommendations that resulted from the FAA/industry forum.[180] The Board also requested that the FAA indicate whether it was able to survey all operators' lubrication practices, as called for in Safety Recommendation A-01-47, and to describe any long-term actions planned as a result of the forum. Pending receipt of the requested information, the Board classified Safety Recommendations A-01-47 and -48 "Open—Acceptable Response."

1.18.4 Alaska Airlines Fleetwide MD-80 Jackscrew Assembly Data and Tracking History

Alaska Airlines and Boeing provided data to the Safety Board so that it could produce a complete inventory of all jackscrew assemblies used in the Alaska Airlines fleet. The data included the tail numbers of the airplanes in which the jackscrew assemblies were installed, information about whether the airplanes were original airline airplanes, the date that the airplanes were delivered to Alaska Airlines, the total flight hours on the jackscrew assemblies, the bench-check end play measurements for the jackscrew assemblies,[181] the date of removal and replacement of the jackscrew assemblies, and the computed wear rate of the jackscrew assemblies.

The data indicated that three jackscrew assemblies, including the accident assembly and the assemblies from N981AS and N982AS, had excessive wear rates. According to maintenance records and information provided by Alaska Airlines, the

---

[180] FSAW 02-02C, "The Potential Adverse Effects of Grease Substitution," (amended on May 22, 2002) stated that presentations at the FAA/industry forum indicated that "there have been no equipment failures attributed to changing greases when lubrication practices (lubrication type, frequency, and technique) are per the original equipment manufacturer's recommendations; and each lubrication point is properly purged" and that "when properly purged, the resulting intermixing of greases does not pose an airworthiness concern. However, the lack of lubrication can result in damage to mechanisms that require lubrication for normal operation."

[181] Most of the end play measurements were taken during bench checks after the jackscrew assemblies were removed. If no bench-check end play measurements could be obtained, the measurement from the most recent on-wing end play check was used.

jackscrew assemblies from N981AS and N982AS received their most recent lubrications at Alaska Airlines' maintenance facility at OAK (during C checks) and the lubrications before that at its maintenance facility at SFO.[182]

N981AS

On February 9, 2000, metal shavings were found in another Alaska Airlines MD-83, N981AS, during a fleet inspection in Portland, Oregon. An on-wing end play check was performed on N981AS, and the end play measurement was 0.055 inch. At the time of the end play check, the airplane had accumulated 10,201 flight hours.

The jackscrew assembly was removed and sent to the Safety Board's Materials Laboratory for examination. Safety Board investigators found metal shavings, some measuring 1/8 to 1/4 inch long, in the area directly beneath the acme screw and in the grease on the acme screw. Several semicircular shavings about 2 1/5 inches long were also found. After the assembly was cleaned, an end play check was performed; the end play measurement was 0.053 inch in a loaded condition.

N982AS

Also on February 9, 2000, metal shavings were found in an Alaska Airlines MD-83, N982AS, during a fleet inspection at SEA. Several on-wing end play checks were performed on N982AS, which had accumulated 9,980 flight hours. At a torque of 275 inch-pounds, the end play measurements were 0.0510, 0.0525, and 0.0525 inch on three successive attempts. At a torque of 250 inch-pounds, the end play measurement was 0.082 inch on three successive attempts.

The jackscrew assembly was removed and sent to the Safety Board's Materials Laboratory for examination. Safety Board investigators found metal shavings, some measuring 1/8 inch long, in the area directly beneath the acme screw and in the grease on the acme screw. A semicircular shaving about 2 1/5 inches long was also found. After the assembly was cleaned, an end play check was performed; the end play measurement was 0.048 inch in a loaded condition.

The data also indicated that, in 1999, Alaska Airlines had removed and replaced the jackscrew assemblies from the following MD-80s in its fleet: N947AS, N975AS, and N951AS.

N947AS

Jackscrew assembly S/N DCA 1935 was removed from N947AS in June 1999 at Alaska Airlines' OAK maintenance facility during a 30,000-flight-hour check because of excessive wear. N947AS was delivered new to Alaska Airlines in December 1990. At the

---

[182] As previously mentioned, the accident jackscrew assembly received its last lubrication at Alaska Airlines' SFO maintenance facility. According to Alaska Airlines, nearly all of its MD-80s had received at least one jackscrew assembly lubrication at its SFO or OAK maintenance facility during the 3 years before the flight 261 accident.

time of the jackscrew assembly removal, N947AS had accumulated 27,860 flight hours (15,906 flight cycles). No end play measurements were recorded. However, a bench end play check conducted at Trig Aerospace (currently Integrated Aerospace) on February 27, 2000, indicated an end play measurement of 0.029 inch.

N975AS

Jackscrew assembly S/N DCA 2272 was removed from N975AS on November 17, 1999, at Alaska Airlines' OAK maintenance facility during a C check because the assembly had a 0.042-inch end play measurement. N975AS was delivered new to Alaska Airlines in May 1994. At the time of the jackscrew assembly removal, N975AS had accumulated 19,960 flight hours (10,786 flight cycles).

Mitchell Aircraft Spares had jackscrew assembly DCA 2272 overhauled by Aerotron AirPower on January 17, 2000. Assembly DCA 2272 was repurchased by Alaska Airlines and sent to Trig Aerospace with an order stating that the unit "needs overhaul to bring to serviceable condition." The assembly was overhauled by Trig Aerospace on February 27, 2000, according to specifications outlined in the Boeing/Douglas Overhaul Maintenance Manual 27-41-1, revision 24, dated April 15, 1998.

N951AS

Jackscrew assembly S/N "AC951" was removed from N951AS on November 26, 1999, by Aviation Management Systems, Phoenix, Arizona, during a 30,000-flight-hour check because the assembly had a 0.045-inch end play measurement. Alaska Airlines purchased N951AS from JetAmerica in October 1987. At the time of the jackscrew assembly removal, N951AS had accumulated 52,894 flight hours (30,284 flight cycles). No maintenance history or information about the jackscrew assembly before or after the acquisition was found.

According to records from a computer-based, statistical component reliability (or alert) program that Alaska Airlines used to track the purchasing of appliances, parts, and components, the airline had not replaced a jackscrew assembly on any airplanes in its MD-80 fleet until 1999.[183] The Alaska Airlines director of reliability and maintenance programs testified at the Safety Board's public hearing that the program, which began in 1995, tracks components "at a rate of 1,000 unit hours in service."

Alaska Airlines' reliability analysis program database indicated that the airline had purchased three replacement jackscrew assemblies in 1999. After replacement of the first jackscrew assembly in June 1999, Alaska Airlines began tracking jackscrew assemblies as rotable components.[184] No earlier maintenance histories (for example, overhaul status, last maintenance, and vendor tear down reports) were available for Alaska Airlines' jackscrew assemblies.

---

[183] Alaska Airlines started operating MD-80s in 1985.

1.18.5 MD-11 Jackscrew Assembly Acme Nut Wear History

In 1995, five operators of McDonnell Douglas MD-11 (MD-11) airplanes reported excessive wear of acme nuts on these airplanes. Subsequent investigations by McDonnell Douglas and operators determined that the excessive wear was a result of improper acme screw thread surface finish.[185] On the basis of this evidence, McDonnell Douglas issued Service Bulletin (SB) MD11-27-067 on July 31, 1997, which provided instructions for measurement, inspection for excessive wear on the actuator assemblies, and repair and replacement, if necessary.

In 1998, another MD-11 operator reported that the acme nut threads on one jackscrew assembly had completely worn away. Boeing determined that this excessive wear was also a result of improper acme screw thread surface finish. The trim system was rendered inoperative by the subsequent failure of the shear pins installed in the drive mechanism that connects the two acme nuts. As a result of this incident, on November 19, 1999, SB MD11-27-067 was escalated to an "Alert" status (ASB MD11-27A-067, revision 5). The summary section of the ASB stated, "If not corrected, this condition, in conjunction with a failure of the opposite side jackscrew assembly, could result in a free horizontal stabilizer that would cause loss of airplane pitch control." On July 30, 1998, the FAA issued an AD that reiterated this concern and required MD-11 operators to comply with ASB MD11-27A-067.

1.18.6 Safety Board Observations of Jackscrew Assembly Lubrications

Safety Board investigators observed lubrications of jackscrew assemblies performed by maintenance personnel from two MD-80 operators and discussed the lubrication procedure with these and other maintenance personnel from these operators. Investigators observed that different methods were used by maintenance personnel to accomplish certain steps in the lubrication procedure, including the manner in which grease was applied to the acme nut fitting and the acme screw[186] and the number of times the trim system was cycled to distribute the grease immediately after its application. In addition, Safety Board investigators who attempted to perform the lubrication procedure noted that, because of the access panel's size, it could be difficult to insert a hand into the access panel and that after a hand was inserted, it blocked their view of the jackscrew assembly, thereby requiring them to accomplish the task primarily by "feel."

---

[184] Before its first jackscrew assembly replacement in June 1999, Alaska Airlines tracked the assemblies as nonstocked items. After the first jackscrew assembly replacement, the assemblies were added to the airline's reliability program rotable component statistical alerting system. Rotables are parts and appliances that are economically repairable and are periodically overhauled to a fully serviceable condition. These parts and appliances have specific requirements to track time and location at a serialized level.

[185] No evidence of improper surface finish was found on the accident acme screw threads. For details about the accident acme screw thread surface finish, see section 1.16.2.2.

[186] The various grease application procedures included the use of a brush, a grease gun nozzle, a glove, a bare hand, or a rag.

A lubrication test performed under Safety Board supervision determined that, when only the grease fitting was lubricated, grease went into the top six or seven acme nut threads before extruding out of the top of the nut. Grease began to extrude out of the top of the acme nut after seven pumps of the grease gun. Lubricating the acme nut through the grease fitting did not provide lubrication to all of the acme screw and nut pressure flanks.

Demonstrations[187] designed to compare the effectiveness of various methods of lubricating the acme screw and nut assembly found that a thorough application of grease onto the entire length of the acme screw, followed by the cycling of the trim several times, maximized the distribution of lubrication on the acme nut threads. Several of the methods observed by or reported to Safety Board investigators did not involve either the application of grease to the entire length of the acme screw or cycling of the trim several times.

1.18.7 End Play Check Anomalies

1.18.7.1 Safety Board End Play Check Observations

Safety Board investigators observed end play checks performed by maintenance personnel from several MD-80 operators, including Alaska Airlines (at its OAK maintenance facility),[188] and observed bench end play checks performed by Integrated Aerospace. Results of hundreds of end play checks submitted to the FAA in accordance with ADs 2000-03-51 and 2000-15-15 were also reviewed.[189]

In its October 2001 safety recommendation letter, the Safety Board noted that the accuracy of the end play results could be affected by deviations in one or more of the following areas: (1) calibration and interpretation of the dial indicator; (2) installation of the dial indicator; (3) application and direction of the specified torque to the restraining fixture;[190] (4) calibration of the torque wrench; (5) fabrication, lubrication, and maintenance of the restraining fixture; (6) rotation of the acme screw within its gearbox during the procedure;[191] and (7) the individual mechanic's knowledge of the procedures.

---

[187] The demonstrations involved the use of acme nuts manufactured from a transparent plastic, which allowed a view of the acme nut threads during lubrication and operation of an acme screw being rotated through the nut under a static load.

[188] In March 2000, Safety Board investigators observed jackscrew assembly end play checks on N932AS, which was undergoing a 17-day C check. The end play check was performed by two sets of mechanics and inspectors.

[189] As stated previously, AD 2000 15-15 requires all U.S. operators of DC-9, MD-80/90, and 717 series airplanes to perform end play checks every 2,000 flight hours and report the end play measurements to the manufacturer after every check. Before the flight 261 accident, operators performed end play checks at various intervals, all but one of which exceeded 2,000 hours. These intervals were based on each airline's maintenance program and were approved by the airline's PMI. For more information about AD 2000-15-15 and the Safety Board's statistical end play data study, see sections 1.18.1 and 1.18.2, respectively.

[190] Torquing the restraining fixture in the wrong direction results in end play measurements of 0.

[191] Rotation of the acme screw can effect the movement of the dial indicator plunger and cause higher end play readings.

During some of the end play check observations, Safety Board investigators evaluated different methods of accomplishing the end play check procedure. Investigators observed that, when the dial indicator was mounted above the acme nut instead of below it (as is called for in the current end play check procedure), the dial indicator was easier to install, its face could be seen more easily (without the need for an inspection mirror, as was the case when it was installed below the acme nut), and the end play measurement could be more easily discerned from the movement of the indicator needle.[192] Investigators also observed that, when the dial indicator was mounted such that the plunger contacted the canted surface of the acme nut stop lug at a skewed angle to the plunger axis, the end play measurement was lower than it was when it contacted a level surface on the acme nut at a right angle to the plunger axis. On April 13 and November 20, 2000, Boeing issued revisions to the end play check procedure.[193]

1.18.7.2 Postaccident Alaska Airlines-Reported Near-Zero End Play Measurements

In March and April 2002, Alaska Airlines removed two jackscrew assemblies from MD-80 airplanes in its fleet following near-zero end play measurements. One jackscrew assembly was removed after a scheduled end play check at Alaska Airlines' maintenance facility at OAK, and the other assembly was removed during a scheduled C check at a contract facility near SEA. Both jackscrew assemblies were removed because of reported end play measurements of 0.001 inch during all iterations of the end play check.[194] Safety Board investigators examined these jackscrew assemblies and determined that the end play measurements were actually higher (within normal and expected limits) and that the 0.001-inch end play measurements were caused by procedural errors made by maintenance personnel while performing the end play check procedure.[195] Specifically, the investigators determined that the maintenance personnel must have torqued the restraining fixture in the wrong direction before measuring the end play.

After these two anomalous end play measurements,[196] the FAA conducted followup inspections at the two facilities. According to the FAA inspectors, all of the restraining fixtures, torque wrenches, and dial indicators used to perform the end play

---

[192] When the dial indicator was mounted below the acme nut, as called for in the current procedure, the dial indicator plunger moved upward as the acme nut moved, causing the needle to move counter-clockwise, or opposite the intended direction. This movement required maintenance personnel to read the dial indicator backward and interpret the total needle movement to obtain the end play measurement.

[193] For more information about Boeing's revisions to the end play check procedure, see section 1.6.3.3.2.1.

[194] The end play check was reiterated about eight times on the jackscrew assembly removed at OAK and about six times on the jackscrew assembly removed at the contract facility near SEA.

[195] A Safety Board examination of the bore of the acme nut from one of the jackscrew assemblies revealed the presence of green grease near the grease fitting entrance and red grease along the bore in all other locations. Grease samples taken from the other jackscrew assembly, from various locations along the acme screw threads immediately above and below the acme nut, exhibited the appearance of a mixture of green and red grease.

[196] Because the manufacturing specifications require a minimum end play of 0.003 inch, the 0.001-inch end play measurements were considered anomalous.

checks were the correct tools and properly calibrated. Further, FAA inspectors indicated that, before these anomalous end play measurements were reported, Alaska Airlines had provided 4 hours of specialized training to all maintenance personnel who perform end play checks at these two facilities. The specialized training focused on how to properly conduct the end play check procedure.

According to Boeing, during the 6-month period from December 2001 to May 2002, six end play measurements of less than 0.004 inch were reported, not including the two 0.001-inch end play measurements reported by Alaska Airlines.

1.18.8 Jackscrew Assembly Overhaul Information

1.18.8.1 Jackscrew Overhaul Specifications and Authority

Airlines and maintenance facilities that have been issued class 1 accessory ratings under 14 CFR 145.31(f) can overhaul jackscrew assemblies without additional FAA authorization. A class 1 accessory rating authorizes the holder to repair and alter "mechanical accessories that depend on friction, hydraulics, mechanical linkage, or pneumatic pressure for operation, including aircraft wheel brakes, mechanically driven pumps, carburetors, aircraft wheel assemblies, shock absorber struts and hydraulic servo units."

According to Boeing, Integrated Aerospace is the only contract facility it uses to perform jackscrew assembly overhauls.[197] When Integrated Aerospace receives a jackscrew assembly with an end play measurement greater than 0.015 inch,[198] it will discard the acme nut and gimbal ring assembly, check that the dimensions of the acme screw are within tolerances, mate a new acme nut and gimbal ring to the acme screw, and assign a new S/N to the nut assembly.[199] Following the buildup of the jackscrew assembly, Boeing requires that the overhauled assembly be functionally tested in a weighted fixture after the assembly is temporarily outfitted to additionally support hardware and electric motors. Boeing also requires new jackscrew assemblies to undergo an operational test in the same test fixture.

---

[197] The instructions and specifications Boeing requires of its maintenance facilities are contained in a service rework drawing. Before the flight 261 accident, information collected from numerous production drawings and proprietary documents formed the basis for the rework instructions. Following the accident, the FAA required Boeing to more formally organize the rework procedures, which resulted in Boeing's publication of the service rework drawing. Integrated Aerospace was the only maintenance facility to possess the drawing. According to Boeing officials, the drawing is proprietary to ensure that the company maintains confidence and control over the precise machining tolerances required.

[198] Jackscrew assemblies may also be overhauled because of excessive free play or damage to the lower mechanical stop because of an electric shutoff failure or misrigging. Not all jackscrew assemblies submitted for overhaul have a high degree of acme nut thread wear.

[199] According to Boeing, the installation of the gimbal ring supports involves precision machining at tolerances as small as five ten-thousandths of an inch to ensure that the alignment of the fastener bores within the gimbal ring supports, the gimbal rings, and the acme nut are adequate to maintain the integrity of the dual load path design.

1.18.8.2 Review of the Boeing DC-9 Overhaul Maintenance Manual

The Safety Board reviewed the section of the Boeing DC-9 Overhaul Maintenance Manual (in 27-41-1, revision 24, dated April 15, 1998, which was not required to be approved by the FAA) titled, "Horizontal Stabilizer Actuator Control Installation." The following observations were noted:

- The manual contained no requirements to functionally test overhauled jackscrew assemblies in a weighted fixture after the assembly is temporarily outfitted to additionally support hardware and electric motors nor to undergo an operational test in the same test fixture, as was required by maintenance facilities per Boeing's service rework drawing.

- The manual contained no requirement for the use of work cards documenting the tasks for overhaul.

- The manual provided no procedures nor equipment listings for bench-check end play measurements.

- The manual contained no requirement to document the bench-check end play measurement after overhaul nor to ensure that the receiver of the overhauled unit was provided with the measurement.

- The manual contained a requirement that, if the axial end play measurement exceeded 0.040 inch, the "nut assembly" (defined in the manual as "the acme screw, acme nut, and gimbal ring support assembly") must be replaced. The manual stipulated that the replacement nut assembly could be new (purchased from Boeing with an end play measurement of between 0.003 and 0.010 inch) or used (with an end play measurement of 0.040 inch or lower). The manual also stated that the overhaul shop could send nut assemblies to Boeing for overhaul.

- The manual contained instructions to check the jackscrew assembly on-wing every 1,000 flight hours if the measured bench-check end play measurement was "between 0.0340 and 0.0390 inch." No such requirement exists in the airplane maintenance manual, which is used by operators.

- The manual contained no procedures or equipment listings pertaining to the surface finish of the acme screw threads. No mention was made of the importance of proper acme screw thread surface finish.

- The manual provided no detail on how to apply grease to the jackscrew assembly.

- The manual did not specify which type of grease should be used to lubricate the acme screw and nut or other assembly components, including the torque tube and gearbox.

- The manual did not specify how to pack the jackscrew assembly for immediate shipping after overhaul; protective packaging instructions were given only for units going into "storage."

1.18.8.3 Safety Board Maintenance Facility Observations

Safety Board investigators visited Integrated Aerospace and three other FAA Part 145 maintenance facilities that overhaul jackscrew assemblies. The visits revealed that two of the three maintenance facilities had self-imposed end play measurement limits at which the acme nut had to be replaced. One maintenance facility's end play measurement limit was 0.025 inch; the other maintenance facility's limit was 0.030 inch. The third maintenance facility had no self-imposed limit at which the acme nut had to be replaced; however, if the end play measurement was less than 0.040 inch, the maintenance facility would return the overhauled assembly to the customer.

The Safety Board's visits also revealed that the maintenance facilities (except for Integrated Aerospace) performed the end play check using different methods. One maintenance facility used a unique, fabricated, self-contained fixture[200] that exerted a force on the acme nut and eliminated acme screw rotation. The end play measurement was taken by a direct dial indicator reading. Another maintenance facility placed the jackscrew assembly on a wooden cradle and moved the acme nut by hand onto the screw while observing a dial indicator. The last maintenance facility used a magnet to hold the dial indicator in place on the acme screw while the nut was held on a table by hand as the end play was checked manually. Only Integrated Aerospace recorded the received and final end play measurements and used a detailed set of work cards to document each step in the overhaul process. The other maintenance facilities documented only that an assembly had been overhauled and used the DC-9 Overhaul Maintenance Manual. Further, two of the four maintenance facilities (not Integrated Aerospace) used subcontractors to apply the black-oxide coating on the acme screw and to ensure that the acme screw thread surface finish met specifications.

The FAA PMI assigned to one of the maintenance facilities stated that he was not aware that the maintenance facility performed overhauls of DC-9 jackscrew assemblies until the facility informed him that it had received a suspect assembly from a foreign airline. Representatives of the maintenance facility stated that they overhauled about six jackscrew assemblies per year and that their customers included aircraft parts brokers and foreign airlines. Representatives of another maintenance facility stated that the jackscrew assemblies were not typically lubricated after overhaul.[201] Representatives of the last maintenance facility indicated that they did not pack the assembly for storage as specified by the DC-9 Overhaul Maintenance Manual because the overhauled assembly was shipped out immediately and not put into storage.[202] This maintenance facility's operators stated that they did not track assemblies after shipment to determine if overhauled assemblies would be stored for long periods of time by parts brokers or airlines.

---

[200] The fixture was not referenced in the DC-9 Overhaul Maintenance Manual.

[201] The DC-9 Overhaul Maintenance Manual instructions call for lubrication of overhauled units.

[202] The storage procedures in the DC-9 Overhaul Maintenance Manual stipulated that jackscrew assemblies be wrapped in barrier material and heat sealed and that 80 percent of the air be evacuated from the wrapping. The manual contained no instructions for packing jackscrew assemblies for shipping.

Safety Board investigators also visited an aircraft parts broker in southern California on February 24, 2001. The broker was listed on an "ILS" subscription listing of organizations that sell jackscrew assemblies and other aircraft components.[203] Investigators examined a recently overhauled acme screw in stock. The jackscrew assembly was stored in an open cardboard box and was wrapped loosely in plastic. Documents that accompanied the assembly indicated that it had been overhauled by a maintenance facility located in the Los Angeles area in November 2000 because of a broken upper mechanical stop. The documents did not contain end play measurements or information about the overhaul. The assembly had red grease on it.

1.18.8.4 Overhauled Jackscrew Assembly Data From 1996 to 1999

Boeing provided the Safety Board with bench-check end play measurement data of all jackscrew assemblies that had been shipped to Integrated Aerospace for overhaul from 1996 to 1999. A review of the data indicated that, out of a total of 214 jackscrew assemblies, 13 had a recorded bench-check end play measurement of 0.045 inch or greater, and 5 had a recorded bench-check end play measurement of 0.055 inch or greater. One jackscrew assembly had a recorded bench-check end play measurement of 0.075 inch, which was the highest measurement reported to investigators during the investigation.

According to Delta Air Lines representatives, the jackscrew assembly (S/N DCA 2239) with the 0.075-inch end play measurement was originally installed on an MD-88 airplane that was delivered new to Delta Air Lines from McDonnell Douglas on March 17, 1993. An on-wing end play check conducted on July 18, 1999, yielded the 0.075-inch end play measurement, and Delta Air Lines removed the jackscrew assembly for overhaul. At the time of the 0.075-inch end play measurement, the airplane had accumulated 18,417 flight hours and 14,520 cycles with the same jackscrew assembly. Integrated Aerospace received the jackscrew assembly on October 29, 1999, and the company also measured a bench-check end play of 0.075 inch. Integrated Aerospace overhauled the jackscrew assembly (with a new acme nut) and returned it to Delta Air Lines on December 17, 1999. According to Delta Air Lines, no further information related to the maintenance history of this jackscrew assembly was found.

1.18.9 Industrywide Jackscrew Assembly Maintenance Procedures

The Safety Board surveyed all 13 domestic airlines that operated at least 10 DC-9, MD-80/90, and/or 717 airplanes in their fleets. The airlines were asked to report their jackscrew assembly lubrication and end play intervals before the flight 261 accident. In addition, the airlines were asked to report the type of grease and the internally imposed maximum allowable end play limit they used before the accident and the number of jackscrew assemblies they removed in 1998 and 1999. The airlines were also asked to

---

[203] The ILS subscription listing is used by overhaul shops and brokers to advertise overhauled parts for purchase by airlines.

provide the maintenance work cards for both the lubrication and end play check of the jackscrew assembly. A review of the survey results revealed the following pertinent information:

- Alaska Airlines was the only airline that used Aeroshell 33 to lubricate the jackscrew assembly. All of the other operators used an MIL-G-23827-qualified grease, such as Mobilgrease 28, Aeroshell 7, or Aeroshell 22.

- Alaska Airlines had the highest MD-80 jackscrew assembly lubrication interval (approximately 2,550 flight hours).

- Alaska Airlines had the second highest MD-80 end play check interval (approximately 9,550 flight hours). The highest end play check interval at all of the airlines was approximately 10,500 flight hours.

- The lowest MD-80 jackscrew assembly lubrication interval was 600 flight hours.

- The operators' MD-80 end play check intervals varied from 7,000 to 10,500 flight hours.

- The operators' DC-9 jackscrew assembly lubrication intervals varied from 550 to 2,530 flight hours.

- The operators' DC-9 end play check intervals varied from 2,000 to 7,720 flight hours. One operator did not use the end play check and instead had a required a hard-time removal at 15,500 hours.

- All of the MD-90 and 717 operators' jackscrew assembly lubrication intervals were set at 3,600 flight hours.

- All of the MD-90 and 717 operators' end play check intervals were set at 7,200 flight hours.

1.18.10 FAA Commercial Airplane Certification Process Study

In February 2002, the FAA issued a report titled, "Commercial Airplane Certification Process Study – An Evaluation of Selected Aircraft Certification, Operation and Maintenance Processes." The report was based on a study that was led by the FAA and conducted by a team that included representatives from U.S. and non-U.S. airplane manufacturers, the U.S. aviation industry, the National Aeronautics and Space Administration, the Air Line Pilots Association, the DoD, Sandia National Laboratories, the Fed Ex Pilots Association, and an international airworthiness consultant.

The airplane certification process study report stated that the study was conducted because "recent accidents such as the Alaska Airlines MD-80 in January 2000, or the Trans World Airways B-747 in July 1996, have brought into question the adequacy of certain processes related to the certification of the airplane." The report focused on numerous process categories, including human factors in airplane design, operations and maintenance, flight critical systems and structure, safety data management, maintenance/operations coordination, major repairs and modifications, and safety oversight. The report added, "special emphasis was also placed upon analyzing how the various major certification processes in the airplane's life cycle relate to each other, and evaluating the relationship between these certification processes and the maintenance and operating processes being applied in service."

The airplane certification process study report's findings and observations included the following:

- Design techniques, safety assessments, and regulations do not adequately address the subject of human error in design or in operations and maintenance.

- There is no reliable process to ensure that the assumptions made in the safety assessments are valid with respect to operations and maintenance activities and that operators are aware of these assumptions when developing their operations and maintenance procedures. In addition, certification standards may not reflect the actual operating environment.

- Several catastrophic events may have been avoided by use of a more robust design and in-depth safety analysis that challenged the assumptions of the design process. This is particularly important when the catastrophic event is the end-point of a short fault tree (two or three failures) or where very specific human actions are required to prevent an accident.

- Processes for identification of safety critical features of the airplane do not ensure that future alterations, maintenance, repairs, or changes to operational procedures can be made with cognizance of those safety features.

- Multiple FAA-sponsored data collection and analysis programs exist without adequate interdepartmental coordination or executive oversight.

- There is no widely accepted process for analyzing service data or events to identify potential accident precursors.

- Adequate processes do not exist within the FAA or in most segments of the commercial aviation industry to ensure that the lessons learned from specific experiences in airplane design, manufacturing, maintenance, and flight operations are captured permanently and made readily available to the aviation industry. The failure to capture and disseminate lessons learned has allowed airplane accidents to occur for causes similar to those in past accidents.

- There are currently no industry processes or guidance materials available which ensure that (1) safety-related maintenance or operational recommendations developed by the manufacturer are evaluated by the operator for incorporation into their maintenance or operational programs and (2) safety-related maintenance or operational procedures developed or modified by the operator are coordinated with the manufacturer to ensure that they do not compromise the type design safety standard of the airplane and its systems.

- The lack of adequate formal business processes between FAA Aircraft Certification Service and Flight Standards Service limits effective communication and coordination between the two that often results in inadequate communications with the commercial aviation industry.

- Processes to detect and correct errors made by individuals in the design, certification, installation, repair, alteration, and operation of transport-category airplanes are inconsistent, allowing unacceptable errors in critical airworthiness areas.

Referring to the lubrication-related safety recommendations the Safety Board issued on October 1, 2001, the report stated that "while the relevance of mixing grease types in this accident has not been determined, both types of grease were discovered in the jackscrew assembly of the accident aircraft. It appears that the communication between operator, [airplane manufacturer,] and the grease manufacturers was not adequate to ensure that the grease mixing recommendations [by Boeing and the grease manufacturers] were communicated and incorporated."[204]

1.18.11 Fail-Safe Jackscrew Assembly Designs

Representatives of Nook Industries, which manufactures jackscrew assemblies for nonaviation applications, indicated to Safety Board investigators that the jackscrew assemblies it designs for safety-critical functions are equipped with a secondary acme nut. The secondary acme nut, which is also referred to as a "follower nut," is threaded on the acme screw and follows the primary load-bearing acme nut as it rotates longitudinally along the acme screw. Engineers at the National Aeronautics Space Administration's (NASA) John F. Kennedy Space Center have developed a similar mechanism for use in the jackscrew assembly used for the gaseous oxygen vent arm on its space shuttle launch pads. The secondary acme nut in both the Nook and NASA designs does not carry any load unless the primary nut threads shear off. If this occurs, the secondary acme nut assumes the load and allows the system to continue to operate. The addition of the secondary acme nut also provides an alternate method for monitoring end play between the acme screw and nut threads that does not require relieving the load on the primary nut. As the threads of the primary acme nut wear, the distance between the primary and secondary nut decreases; therefore, wear of the primary nut threads, or end play, can be gauged by monitoring changes in this distance.

---

[204] Boeing service letters issued on August 12, 1993, and June 30, 1997, noted that these grease types were incompatible "and should not be mixed."

2. Analysis

2.1 General

The flight crewmembers on Alaska Airlines flight 261 were properly certificated and qualified and had received the training and off-duty time prescribed by Federal regulations. No evidence indicated any preexisting medical or other condition that might have adversely affected the flight crew's performance during the accident flight.

The airplane was dispatched in accordance with Federal Aviation Administration (FAA) regulations and approved Alaska Airlines procedures. The weight and balance of the airplane were within limits for dispatch, takeoff, climb, and cruise.

Daylight visual meteorological conditions prevailed, and weather was not a factor in the accident.

Examination of recovered wreckage and cockpit voice recorder (CVR), flight data recorder (FDR), and air traffic control (ATC) data revealed no evidence of a fire or of impact with birds or any other foreign object.

No evidence indicated that the airplane experienced any preimpact structural or system failures, other than those associated with the longitudinal trim control system, the horizontal stabilizer, and its surrounding structure.[205]

FDR and CVR information indicated that both engines were operating normally before the final dive. Examination of the recovered thrust reversers and FDR data indicated that the thrust reversers were stowed at impact. Examination of the wreckage indicated that the left engine was producing significant rotational energy at impact and that the right engine was producing very little rotational energy at impact. Erratic engine pressure ratio data recorded on the FDR during the final dive were consistent with airflow disturbances associated with extreme angles-of-attack. Such extreme angles-of-attack can cause fuel unporting and can interrupt the airflow into the engine inlet, either of which could lead to an engine flameout. Therefore, the lack of rotation on the right engine was likely the result of an engine flameout during the final dive.

ATC personnel involved with the accident flight were properly certificated and qualified for their assigned duty stations. ATC radar and communications equipment were functioning properly during the flight.

Analysis of the structural damage and FDR data indicated that the airplane was nearly inverted[206] and was in a steep nose-down, right-wing-low attitude when it impacted the water, which is consistent with uncontrolled flight.

---

[205] For an analysis of this failure sequence, see sections 2.2.2 through 2.2.4.

This analysis examines the accident scenario, including the catastrophic failure of the longitudinal trim control system, flight crew decision-making, possible reasons for the excessive acme nut thread wear, the importance of monitoring acme nut thread wear, jackscrew assembly overhaul procedures and facilities, adequacy of Alaska Airlines' maintenance program, FAA oversight, and horizontal stabilizer design issues.

2.2 Accident Sequence

2.2.1 Takeoff and Climbout

FDR data indicated that the accident airplane's longitudinal trim control system was functioning normally during the airplane's descent into Lic Gustavo Diaz Ordaz International Airport (PVR), Puerto Vallarta, Mexico, on the flight just before the accident flight. FDR data indicated that the airplane landed at PVR with a horizontal stabilizer angle of 6° airplane nose up.

FDR data indicated that the accident flight crew had trimmed the airplane to a 7° airplane-nose-up position before takeoff from PVR. After the airplane took off about 1337, FDR data indicated that the horizontal stabilizer moved at the normal primary trim motor rate of 1/3° per second from 7° to 2° airplane nose up. Thereafter, as the airplane continued to climb from 6,200 feet to about 23,400 feet, the FDR recorded horizontal stabilizer movement at the normal alternate trim motor rate of 1/10° per second from 2° airplane nose up to 0.4° airplane nose down. Operation of the alternate trim motor during this period is consistent with pitch control being commanded by the autopilot, which was activated at 1340:12 while the airplane was climbing through 6,200 feet. On the basis of these data, the Safety Board concludes that the longitudinal trim control system on the accident airplane was functioning normally during the initial phase of the accident flight.

2.2.2 Jamming of the Horizontal Stabilizer

FDR data indicated that the horizontal stabilizer's last movement during the climbout was to 0.4° airplane nose down at 1349:51 as the airplane was climbing through 23,400 feet at 331 knots indicated airspeed (KIAS). After this, no horizontal stabilizer movement was recorded on the FDR until the airplane's initial dive 2 hours and 20 minutes later. This cessation of horizontal stabilizer movement is not consistent with a typical MD-80 climb profile, which normally would require additional stabilizer movements to maintain trim. Thus, as the accident airplane continued to climb above 23,400 feet without any horizontal stabilizer movement, the autopilot would have attempted to achieve trim by continuing to add elevator input to compensate for the lack of movement of the horizontal stabilizer.

---

[206] The largest intact sections of the recovered fuselage were from below the airplane's floor line, indicating that the airplane was rolling toward inverted at impact.

After the last horizontal stabilizer movement, the airplane continued to climb at 330 KIAS. The airplane began to level off at approximately 26,000 feet and slowed to 320 KIAS. At 1351:51, the airplane began to climb again. By 1352:50, the airplane had slowed to about 285 knots as it climbed to approximately 28,600 feet and then leveled off.

At 1353:12, when the airplane was at approximately 28,557 feet and 296 KIAS, the autopilot was disconnected, most likely by the flight crew in response to the trim annunciator warning light, which would have illuminated to indicate the airplane's out-of-trim condition.[207] In a typical climb profile, a reduction in speed during the ascent would result in an airplane-nose-up stabilizer movement to trim the airplane. At the time of the autopilot disconnect, the airplane was still trimmed for an airspeed of 331 knots (the airspeed when the last horizontal stabilizer movement occurred), 35 knots faster than the airspeed at that time. After the autopilot was disconnected, the elevators were deflected in the airplane-nose-up direction from 0° to approximately -1.5°.

For almost 7 minutes after the autopilot disconnect, the airplane continued to climb at a much slower rate than before the disconnect, reaching about 31,050 feet by 1400:00. During this part of the ascent, the elevators were deflected in the airplane-nose-up direction between -1° and -3°, which, according to airplane performance calculations, would have required up to 50 pounds of pulling force on the control column(s).[208] The calculations further indicated that, after leveling off at 31,050 feet, only about 30 pounds of pulling force would have been required to maintain level flight for the next 24 minutes while the airplane was flying at 280 KIAS. This was followed by an increase in airspeed[209] beginning at 1424:30 that, along with the decreased weight of the airplane because of fuel usage, would have eased the amount of pulling force needed to maintain level flight to roughly 10 pounds until the autopilot was switched back on about 1547.

After the autopilot disconnect at 1353:12, the flight crew would have noticed that the airplane was mistrimmed because significant pilot control forces would have initially been needed to maintain the desired climb and cruise configuration, and the flight crew would have likely attempted to correct the problem by manually activating the primary or alternate trim systems to move the horizontal stabilizer. However, the horizontal stabilizer remained at 0.4° airplane nose down, indicating that the flight crew was unable to manually command movement of the horizontal stabilizer. During discussions with Alaska Airlines maintenance personnel at Los Angeles International Airport, (LAX), Los Angeles, California, later in the flight, the captain indicated that the flight crew had tried numerous troubleshooting strategies to move the horizontal stabilizer, including checking the circuit breakers and using the control wheel trim switches and longitudinal trim handles on the center pedestal separately and together.[210] The captain told the maintenance

---

[207] The airplane's speed decreased (from 331 to 296 KIAS) during the minutes preceding the autopilot disconnect, indicating that the airplane was mistrimmed.

[208] As stated previously, all estimates of pulling force in this report refer to the combined column forces from both the captain's and the first officer's control columns. It is not known whether these forces were applied by the captain, the first officer, or both.

[209] Although the flight plan called for a cruise speed of 283 knots calibrated airspeed, FDR data indicated that the airplane's cruise speed increased, starting at 1424:30, eventually reaching 301 KIAS.

personnel, "it appears to be jammed...the whole thing, it spikes out when we use the primary, we get AC load that tells me the motor's tryin to run but the brake won't move it. when we use the alternate, nothing happens."[211]

In sum, the Safety Board concludes that the horizontal stabilizer stopped responding to autopilot and pilot commands after the airplane passed through 23,400 feet. The pilots recognized that the longitudinal trim control system was jammed, but neither they nor the Alaska Airlines maintenance personnel could determine the cause of the jam.

2.2.2.1 Cause of the Jam

As previously mentioned, the horizontal stabilizer's jackscrew assembly includes an acme screw and nut, both of which have two threads. After the accident, severely worn and sheared remnants of the acme nut threads were recovered wrapped around the acme screw. The condition of the recovered acme nut thread remnants indicated that approximately 90 percent of the thread thickness had worn away before the remainder of the threads sheared off.[212] In comparison, a jackscrew assembly with the maximum amount of wear permitted in service, as indicated by an end play measurement of 0.040 inch (that is, 0.030 to 0.037 inch of wear), would only have about 22 percent of the thread thickness worn away.[213]

The interior of the recovered acme nut contained no intact threads, although there were small ridges where the worn nut thread remnants had previously been attached. The inner diameter of those ridges was almost identical to the outer diameter of the acme screw threads, which is consistent with the worn acme nut threads having been sheared off by the acme screw threads during airplane-nose-down trimming movements.

Each of the acme nut threads makes 16 360° spiral revolutions, for a total of 32 360° revolutions. Calculations indicated that loads of more than 190,000 pounds would be necessary to simultaneously shear all 32 360° spiral revolutions of the acme nut threads if they were worn to the same degree as the accident acme nut thread remnants. However, airplane performance data determined that the flight loads on the jackscrew assembly were only 4,000 to 5,000 pounds during the climbout from PVR. Therefore, it is very unlikely that the worn acme nut thread spiral revolutions were simultaneously sheared along their entire length.

---

[210] When Alaska Airlines maintenance personnel asked the captain whether the pilots had tried "the suitcase handles [colloquial term for longitudinal trim handles] and the pickle switches [colloquial term for control wheel trim switches]" in their attempts to troubleshoot the horizontal stabilizer trim problem, the captain replied that they had "tried everything together." When the maintenance technician subsequently asked again whether the flight crew had tried the control wheel trim switches and longitudinal trim handles, the captain replied, "yea we tried just about every iteration."

[211] The alternate trim motor uses less power than the primary trim motor; therefore, activation of that system does not normally produce enough electrical current to cause any appreciable movement of the needle on the AC load meter.

[212] For more information about the condition of the recovered acme nut threads, see section 1.16.3.6.

[213] This calculation was based on a comparison with the manufacturing specifications for the thickness of the threads at the major diameter (0.15 inch).

The calculations further indicated that approximately 6,000 to 7,000 pounds would be necessary to shear one 360° spiral revolution of an acme nut thread if it were worn to the same degree as the accident acme nut thread remnants. However, the Safety Board's finite element analysis (FEA) study showed that, at high levels of thread wear, such as those of the accident acme nut threads, bending stresses develop that are significantly larger than the stresses caused by direct contact pressure. The study showed that the bending deformations resulting from a 2,000-pound load on a single highly worn thread spiral caused a drastic change in the thread's behavior, indicating that a structural failure, rather than shearing failure, is possible.

Further, the results of the Safety Board's and Boeing's FEA of the stress and deformation in the acme nut threads under loading indicated that the wear process is an incremental one in which the loads and resulting wear shift and move among all of the 32 spiral revolutions in the acme nut.[214] Thus, an incremental bending-induced shear fracture of the threads most likely propagated along the thread spiral, eventually leading to complete failure. Figure 20 depicts the stages of acme nut thread wear to the point of fracture.

The Safety Board concludes that the worn threads inside the horizontal stabilizer acme nut were incrementally sheared off by the acme screw and were completely sheared off during the accident flight. The Safety Board further concludes that, as the airplane passed through 23,400 feet, the acme screw and nut jammed, preventing further movement of the horizontal stabilizer until the initial dive. Safety Board investigators considered numerous factors that might have played a role in the jamming of the acme screw to the nut, including bending of the worn threads before or during the shearing, distortion of the remnants after they were sheared,[215] and loads resulting from the acme screw threads pulling upward across the remaining ridges after the shearing;[216] however, the Board could not determine the exact cause of the jam.

---

[214] For more information about these studies, see section 1.16.4.

[215] Because the thinnest part of the sheared thread remnants was too large to fit in the space between the crowns of the acme screw threads and the stripped surface of the nut, the remnants themselves probably played only a minor role, if any, in the creation of the jam.

[216] It was theorized that, as the last of the acme nut thread remnants shear fractured and the screw threads tried to override the resulting ridges in the nut as the screw was pulled upward by aerodynamic loads and rotated circumferentially by the primary trim motor, the combined pinching and friction loads became greater than the combined aerodynamic and primary trim motor forces.

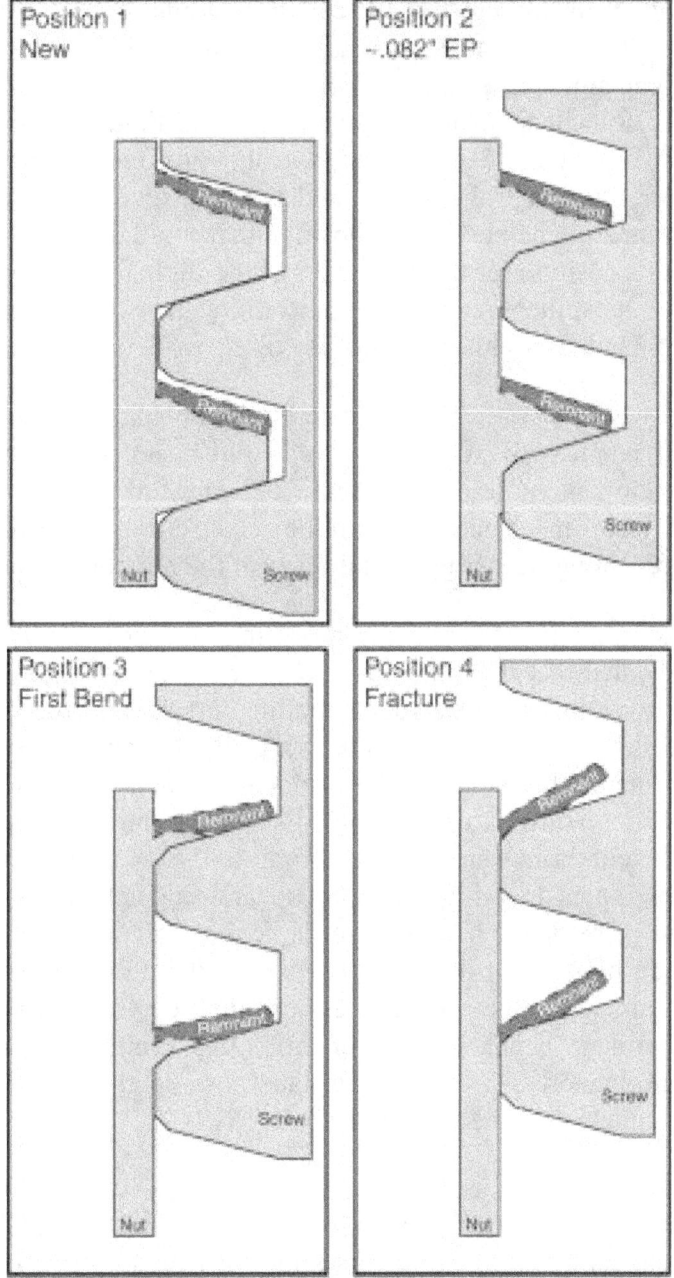

Figure 20. A graphical depiction of the stages of acme nut thread wear to the point of fracture.

2.2.3 Release of the Jam and the Initial Dive

At 1608:59, following discussions with Alaska Airlines maintenance personnel about whether the fight crew had tried the longitudinal trim handles and the control wheel trim switches, the captain told the first officer, "I'm going to click it off you got it [the airplane]...lets do that." At this time, the airplane was cruising at 31,050 feet. At 1609:14.8, the captain stated, "this'll click it off," and, about 1609:16, the autopilot parameter on the FDR switched from "engaged" to "off." FDR and airplane performance data indicated that during the 3 to 4 seconds following the autopilot disconnect, the horizontal stabilizer traveled from its jammed position of 0.4° airplane nose down to beyond the maximum airplane-nose-down position that could be recorded by the FDR and remained beyond this position for the remainder of the flight.[217] Analysis of the FDR data showed that the trim movement rate was not uniform, which was inconsistent with normal trim operation commanded by either the primary or alternate trim motor.

As the jam was overcome, the acme screw was being pulled upward through the acme nut by aerodynamic loads, causing upward movement of the horizontal stabilizer, resulting in greater airplane-nose-down motion. This upward pulling motion would have continued until the lower mechanical stop on the acme screw contacted the lower surface of the acme nut, preventing further upward motion of the horizontal stabilizer.

Boeing engineering data, computer modeling, and ground demonstrations on an actual airplane indicated that the maximum possible horizontal stabilizer position when the acme screw lower mechanical stop is pulling up against the acme nut is about 3.1° airplane nose down. This is consistent with airplane performance studies of the airplane's movements at this point in the accident sequence that showed the airplane beginning to pitch nose down to nearly -8°. FDR data showed that during this 3- to 4-second period, the elevators were deflected from -0.4° to -4.0°, indicating that the flight crew made control inputs to bring the nose up and counteract the movement of the horizontal stabilizer. The speed brakes were deployed at 1609:43, 27 seconds after the autopilot disconnect, most likely in an attempt by the flight crew to retard the airplane's rapidly increasing airspeed. The speedbrakes remained deployed for the next 1 minute 30 seconds. The airplane began to recover from this initial dive and leveled off at about 1611 at approximately 24,000 feet. Airplane-nose-up elevator inputs continued for about the next 8 minutes (until the final dive about 1619).

Safety Board investigators attempted to determine what caused the horizontal stabilizer to move immediately after the autopilot disconnect. Normally, the autopilot is disconnected by the pilot flying using the autopilot disconnect switch located on the outboard side of the control wheel. However, the autopilot will also disconnect when the primary trim control system is activated by either the control wheel trim switches or the longitudinal trim handles on the center pedestal. After the initial dive, the captain told the

---

[217] The maximum airplane-nose-down position of the horizontal stabilizer during normal operation is 2.1°. The FDR recorded airplane-nose-down positions up to about 2.5°. Airplane performance data indicated that the stabilizer position had to have been beyond 2.5° airplane nose down to result in the airplane's pitch movements during and after the initial dive.

Alaska Airlines LAX maintenance technician, "we did both the pickle switch [control wheel trim switches] and the suitcase handles [longitudinal trim handles] and it ran away full nose trim down [sic]." Based on this comment and those made by the Alaska Airlines LAX maintenance personnel immediately preceding the autopilot disconnect, it appears that the captain disconnected the autopilot when he activated the primary trim control system by using either the control wheel trim switches, the longitudinal trim handles, or both.

Consequently, although operation of the primary trim motor as part of troubleshooting attempts earlier in the flight did not release the jam, the torque created by the primary trim motor when the captain activated the primary trim system at 1609:16 apparently provided enough force to overcome the jam between the acme nut and screw. Over the next 3 to 4 seconds, the increasing angle-of-attack of the horizontal stabilizer and the increased elevator deflections would have in turn increased the tension loads on the acme screw, contributing to the motion of the acme screw upward through the acme nut.

Therefore, the Safety Board concludes that the accident airplane's initial dive from 31,050 feet began when the jam between the acme screw and nut was overcome as a result of the operation of the primary trim motor. Release of the jam allowed the acme screw to pull up through the acme nut, causing the horizontal stabilizer leading edge to move upward, thus causing the airplane to pitch rapidly downward.

Markings on the recovered jackscrew assembly lower mechanical stop and the lower surface of the acme nut indicated that the nut repeatedly contacted the stop in various positions and that the acme screw and attached lower mechanical stop had rotated in both directions while the stop was in contact with the nut. Further, the rotationally sheared splines on the lower mechanical stop indicated that the acme screw rotated within the lower mechanical stop after the stop contacted the lower surface of the acme nut. Metallurgical analysis determined that the marks on the lower mechanical stop could have been created by rotation of the acme screw only after complete loss of the acme nut threads. Circumferential marks were found in the acme nut bore, also indicating acme screw rotation after the threads had sheared off. Such rotation of the acme screw could only be caused by operation of a trim motor. Therefore, the Safety Board concludes that the acme screw did not completely separate from the acme nut during the initial dive because the screw's lower mechanical stop was restrained by the lower surface of the acme nut until just before the second and final dive about 10 minutes later.

2.2.4 The Second and Final Dive

After the flight crew recovered from the initial dive and leveled off at approximately 18,000 feet, the captain instructed the first officer to deploy the slats and flaps (at 1617:54 and 1618:05, respectively), most likely to evaluate flight control pressures. FDR data indicated that the airspeed decreased about 8 knots and that the airplane remained stable and controllable following the slat and flap deployment. At 1618:26, the captain asked the first officer to retract the slats and flaps for unknown reasons. During the 50 seconds after the slats and flaps were retracted, the airplane accelerated from about 245 to 270 KIAS.

Beginning at 1619:09, the FDR recorded small oscillations in elevator angle, pitch angle, and vertical acceleration values for about 15 seconds. At 1619:21, the CVR recorded a series of faint thumps.[218] Although the horizontal stabilizer position had already exceeded the maximum position capable of being sensed and recorded by the FDR, the FDR nonetheless indicated an increase in the horizontal stabilizer position at the time of the faint thump.[219] This increase was the first horizontal stabilizer movement since the one that precipitated the initial dive about 10 minutes earlier. The FDR contained no direct data about the actual position to which the horizontal stabilizer moved at this time. However, Boeing engineering design data and computer modeling of the horizontal and vertical stabilizers indicated that if the acme screw was free to travel through the acme nut, the horizontal stabilizer would move from the 3.1° airplane-nose-down position it was being held at by the lower mechanical stop and would contact the vertical stabilizer tip fairing brackets at a 3.6° airplane-nose-down angle. Structural examinations determined that this additional movement of the horizontal stabilizer could only result from a fracture of the acme screw torque tube.

Metallurgical analysis of the recovered section of the fractured torque tube showed that the final separation initiated from an area of low-cycle fatigue cracking. On the basis of an evaluation of the static loading conditions and performance data, the Safety Board determined that the torque tube was capable of sustaining a single application of the maximum load applied to the torque tube between the initial and final dives.[220] However, low-cycle fatigue tests and analysis, performance data, and FEA established that the magnitude and frequency of the loads on the torque tube after complete shearing of the acme nut threads, in combination with an offset loading condition,[221] would be sufficient to initiate and propagate low-cycle fatigue cracking in the torque tube. The low-cycle fatigue cracking severely reduced the load capability of the torque tube during the 10-minute period between the initial and second and final dive, causing it to fracture at loads considerably less than it was capable of sustaining if it did not have fatigue damage.

At 1619:29, the captain ordered redeployment of the slats and flaps. At 1619:35, the flaps were transitioning from 7° to 11°. At 1619:36.6, the CVR recorded the sound of an "extremely loud noise." Immediately thereafter, the airplane began its final dive. At the time of this pitchover, radar detected several small primary returns, consistent with parts of the vertical stabilizer's tip fairing being torn from the airplane as the fairing brackets

---

[218] According to the Safety Board's sound spectrum study, the "faint thump" noted on the CVR transcript at 1619:21 was actually a series of thumps.

[219] After the sound of the faint thump, the elevator deflections increased from -9° to -13°, indicating the pilots were making additional airplane-nose-up elevator inputs to maintain level flight with the increased airplane-nose-down horizontal stabilizer angle.

[220] For more information about these loads, see section 1.16.5

[221] The torque tube was subjected to a combination of bending and tension as a result of an offset loading condition created in the torque tube because the stop lugs on the lower mechanical stop and the acme screw prevented application of equal load around the axis of the tube.

broke,[222] which would have allowed the horizontal stabilizer to move well beyond the 3.6° airplane-nose-down position it was being held at by the brackets. Therefore, the extremely loud noise recorded on the CVR at 1619:36.6 was likely made as the fairing brackets failed and caused the loss of the tip fairing and structural deformation of the tail under flight loads, resulting in local aerodynamic disturbances.[223]

Valid airplane performance data do not exist for horizontal stabilizer positions beyond 14° airplane nose down. Although the exact horizontal stabilizer angle required to produce the final pitchover could not be determined, the Safety Board's airplane performance studies determined that the airplane's pitch rate during the final dive could only have resulted from a rapid increase to a stabilizer angle significantly beyond 14° airplane nose down.[224]

In sum, the Safety Board concludes that the cause of the final dive was the low-cycle fatigue fracture of the torque tube, followed by the failure of the vertical stabilizer tip fairing brackets, which allowed the horizontal stabilizer leading edge to move upward significantly beyond what is permitted by a normally operating jackscrew assembly. The resulting upward movement of the horizontal stabilizer leading edge created an excessive upward aerodynamic tail load, which caused an uncontrollable downward pitching of the airplane from which recovery was not possible.

2.2.5 Flight Crew Decision-Making

2.2.5.1 Decision to Continue Flying Rather than Return to PVR

Safety Board investigators considered several reasons that might explain the captain's decision not to return immediately to PVR after he experienced problems with the horizontal stabilizer trim system during the climbout from PVR.

Neither the Alaska Airlines MD-80 Quick Reference Handbook (QRH) *Stabilizer Inoperative* checklist nor the company's QRH *Runaway Stabilizer* emergency checklist required landing at the nearest suitable airport if corrective actions were not successful. These checklist procedures were the only stabilizer-related checklist procedures contained in the QRH, and the flight crew most likely followed these checklist procedures in their initial attempts to correct the airplane's jammed stabilizer.

---

[222] The Safety Board notes that the fairing brackets were not intended to carry the load of the horizontal stabilizer.

[223] As discussed in section 1.11.1.1, sound tests indicated that neither this sound nor any of the other sounds recorded by the accident airplane's CVR were similar to sounds associated with the failure of the torque tube. The investigation could not determine exactly what caused the sound.

[224] Structural analysis indicated that the horizontal stabilizer must travel to a position of greater than 15° airplane nose down to allow the acme screw to completely exit the top of the acme nut, as it did during the final dive. Symmetrical structural damage was found on the tail, indicating that the horizontal stabilizer's hinge fittings contacted the vertical stabilizer's rear spar, which would occur at a horizontal stabilizer position of 16.2° airplane nose down.

The airplane's takeoff weight of 136,513 pounds was well below the takeoff and climb limits for the departure runway, but it exceeded the airplane's maximum landing weight of 130,000 pounds. Because the airplane did not have an in-flight fuel dumping system, the airplane would have had to remain in flight for about 45 minutes after takeoff until enough fuel had burned to reduce the airplane's weight by the 6,500 pounds needed to reach the airplane's maximum landing weight. A return to PVR to execute an overweight landing would have required higher-than-normal approach speeds for landing and would have created additional workload and risk. An overweight landing at PVR would have been appropriate if the flight crew had realized the potentially catastrophic nature of the trim anomaly. However, in light of the airplane's handling characteristics from the time of the initial detection of a problem to the initial dive, the flight crew would not have been aware that they were experiencing a progressive, and ultimately catastrophic, failure of the horizontal stabilizer trim system.

The flight crew would have been aware that Alaska Airlines' dispatch and maintenance control in Seattle-Tacoma International Airport (SEA), Seattle, Washington, and LAX could be contacted by radio (via ground-based repeater stations) when the airplane neared the United States. However, even though the last horizontal stabilizer trimming movement was recorded by the FDR about 1349:51, the flight crew did not contact Alaska Airlines' maintenance until shortly before the beginning of the CVR transcript about 1549,[225] which suggests that control problems caused by the jammed horizontal stabilizer remained manageable for some time.[226] Further, as previously mentioned, the positive aerodynamic effects of the higher cruise airspeed and fuel burn would have reduced the necessary flight control pressures to roughly 10 pounds and made the airplane easier to control. Therefore, the Safety Board concludes that, in light of the absence of a checklist requirement to land as soon as possible and the circumstances confronting the flight crew, the flight crew's decision not to return to PVR immediately after recognizing the horizontal stabilizer trim system malfunction was understandable.

Although they elected not to return to PVR, later in the flight the flight crew decided to divert to LAX, rather than continue to San Francisco International Airport (SFO), San Francisco, California, where the flight was originally scheduled to make an intermediate stop before continuing to SEA. Comments recorded by the CVR indicated that the flight crew may have felt pressure from Alaska Airlines dispatch personnel to land in SFO.[227] However, after discussing the malfunctioning trim system and current and expected weather conditions at SFO and LAX with Alaska Airlines dispatch and maintenance personnel, the captain decided to land at LAX rather than continue to SFO.

---

[225] The FDR indicates that a continuing series of radio transmissions began over the very-high frequency 2 channel (used for all non-ATC radio transmissions) at 1521. These transmissions continued until the end of the FDR recording, which suggests that the flight crew began contacting Alaska Airlines maintenance personnel at this time.

[226] The Safety Board notes that the 30-minute CVR recording did not capture the flight crew's earlier troubleshooting efforts or the beginning of the flight crew's discussions with maintenance personnel. A longer CVR recording that captured these events would have aided in this investigation. In Safety Recommendation A-99-16, the Board recommended that CVRs on all airplanes required to carry both a CVR and an FDR be capable of recording the last 2 hours of audio. This safety recommendation is currently classified "Open—Unacceptable Response."

The decision to divert to LAX was apparently based on several factors, including more favorable wind conditions at LAX (compared to a direct crosswind at SFO) that would reduce the airplane's ground speed on approach and landing[228] and the captain's concern, expressed to Alaska Airlines dispatch personnel, about "overflying suitable airports." The Safety Board concludes that the flight crew's decision to divert the flight to LAX rather than continue to SFO as originally planned was prudent and appropriate. Further, the Safety Board concludes that Alaska Airlines dispatch personnel appear to have attempted to influence the flight crew to continue to SFO instead of diverting to LAX.

2.2.5.2 Use of the Autopilot

After the flight crew had manually flown the airplane for almost 2 hours, the autopilot was engaged about 1547, disengaged at 1549:56, and re-engaged 19 seconds later. The autopilot remained engaged until 1609:16, just before the initial dive. No discussion on the CVR (which began at 1549) indicated why the autopilot was engaged in either instance. As discussed previously, the increased airspeed and reduced weight would have brought the airplane closer to a trimmed condition, thus allowing autopilot engagement to maintain altitude and heading while further troubleshooting was attempted. However, Alaska Airlines' MD-80 QRH *Stabilizer Inoperative* checklist states, "do not use autopilot" if both trim systems are inoperative. In light of the autopilot's inability to maintain trim (using the alternate trim motor) during the climbout from PVR and the flight crew's subsequent unsuccessful attempts to manually activate the primary trim control system (using the primary trim motor), the flight crew should have known that both the alternate and the primary trim control systems were inoperative. Thus, the flight crew's use of the autopilot was contrary to company procedures.

Because the AC load meter registered electrical spikes when the crew attempted to activate the primary trim system, the flight crew should have realized that the primary trim motor was operational and that the system was jammed beyond the trim motor's capability. Further, engagement of the autopilot, which would have been making automatic elevator corrections to the airplane's mistrimmed condition, masked the airplane's true condition from the flight crew. If the autopilot had subsequently been disconnected without one of the flight crewmembers holding the control wheel and making immediate corrective inputs, the airplane's out-of-trim condition would have resulted in a severe pitch maneuver immediately after the autopilot disconnect. Therefore,

---

[227] At 1552:02, after the captain had stated his intention to divert to LAX, Alaska Airlines dispatch personnel cautioned that if the flight landed at LAX rather than SFO, "we'll be looking at probably an hour to an hour and a half [before the airplane could depart again] we have a major flow program going right now." At 1552:41, the captain responded, "I really didn't want to hear about the flow being the reason you're calling us cause I'm concerned about overflying suitable airports." At 1555:00, the captain commented to a flight attendant, "it just blows me away they think we're gonna land, they're gonna fix it, now they're worried about the flow, I'm sorry this airplane's [not] gonna go anywhere for a while…so you know." After a flight attendant replied, "so they're trying to put the pressure on you," the captain stated, "well no, yea."

[228] According to Alaska Airlines' MD-80 QRH S*tabilizer Inoperative* checklist, 15° of flaps would have been the appropriate flap setting for the accident airplane's approach and landing with a jammed stabilizer; this reduced flap setting would have increased approach speeds and required a corresponding increase in the amount of runway needed during landing.

the Safety Board concludes that the flight crew's use of the autopilot while the horizontal stabilizer was jammed was not appropriate.[229]

2.2.5.3 Configuration Changes

At 1615:56, after recovery from the initial dive, the captain told the air traffic controller that he wanted to "change my configuration, make sure I can control the jet and I'd like to do that out here over the bay if I may." As previously mentioned, the captain then ordered extension of the slats at 1617:54 and the flaps at 1618:05. The captain did not brief the first officer about what to expect or what to do if these configuration changes resulted in excessive flight control pressures or loss of control of the airplane. Further, the captain did not specify that the flaps should be extended at a slower-than-normal rate, which would have been a prudent precaution to minimize the possibility of the configuration change causing abrupt airplane movements that could be difficult to control. Nevertheless, at 1618:17, after the slats and flaps were extended, the captain noted that the airplane was "pretty stable right here." The captain added that the airspeed needed to decrease to 180 KIAS (the airplane was then at 250 KIAS). Nine seconds later, at 1618:26, the captain ordered retraction of the slats and flaps, and the airspeed began to subsequently increase. It was not clear from the CVR recording why the captain ordered retraction of the slats and flaps and allowed the airspeed to increase nor did the CVR recording indicate any discussion about the possible effects of the slat and flap extension.

The Safety Board notes that an airplane with flight control problems should be handled in a slow and methodical manner and that any configuration that would aid a landing should be maintained if possible. On the basis of the captain's comment, the airplane was stable after the slat and flap extension at 1618:05. This configuration would have aided the approach and landing process. The Safety Board concludes that flight crews dealing with an in-flight control problem should maintain any configuration change that would aid in accomplishing a safe approach and landing, unless that configuration change adversely affects the airplane's controllability.

2.2.5.4 Activation of the Primary Trim Motor

At 1618:49, after the slats and flaps were retracted, the captain stated that he wanted to "get the nose up…and then let the nose fall through and see if we can stab it when it's unloaded." The first officer responded, "you mean use this again? I don't think we should…if it can fly." These statements suggest that the captain may have been indicating his intention to retry the primary trim system after reducing aerodynamic forces on the horizontal stabilizer. However, after the first officer's statement at 1619:14, "I think if it's controllable, we oughta just try to land it," the captain abandoned his plan and responded, "ok let's head for LA."

---

[229] The Safety Board also notes that the captain's disconnection of the autopilot when he was the pilot not flying was also contrary to standard industry procedures. Normally, the pilot flying would use the autopilot disconnect switch on the control column and then assess the airplane's controllability.

The Safety Board notes that the earlier repeated attempts to activate the primary trim system went well beyond what was called for in Alaska Airlines' checklist procedures and ultimately precipitated the release of the jam of the acme screw and nut, resulting in the lower mechanical stop impacting the bottom of the nut. An additional attempt to use the trim switches at this point would not have been prudent. The severity of the initial dive changed the situation aboard the airplane to an emergency, which required a more deliberate and cautious approach.

The Safety Board recognizes that, from an operational perspective, the flight crew could not have known the extent of airplane damage. Although flight crews are trained in jammed stabilizer and runaway stabilizer scenarios, the loss of acme nut and screw engagement exceeded any events anticipated in emergency training scenarios, and the flight crew was not trained to devise or execute appropriate configurations and procedures to minimize further damage to the airplane or to prevent the accident. However, the flight crew's earlier attempts to activate the trim motor and configuration changes may have worsened the situation. As previously discussed, the captain's activation of the primary trim motor at 1609:16 precipitated the release of the jam and the initiation of the initial dive. However, it was not clear how many times previous to that the flight crew activated the primary trim motor nor was it clear whether or to what extent the prior activations hastened the release of the jam. Therefore, the Board could not determine the extent to which the activation of the primary trim motor played a role in causing or contributing to the accident.

2.2.5.5 Adequacy of Current Guidance

The Safety Board notes that after the flight 261 accident, Boeing issued a flight operations bulletin outlining procedures to be followed in the event of an inoperative or malfunctioning horizontal stabilizer trim system. The bulletin advised flight crews to

> complete the flight crew operating manual (FCOM) checklist(s). Do not attempt additional actions beyond that contained in the checklist(s). If completing the checklist procedures does not result in operable trim system, consider landing at the nearest suitable airport.

The Safety Board agrees that this advice is generally appropriate. However, the Board does not agree that the flight crew should merely "consider" landing at the nearest suitable airport if accomplishing the checklist items does not result in an operational trim system. In such a case, the flight crew should always land at the nearest suitable airport as expeditiously and safely as possible. Further, the bulletin provides additional information regarding the possibility that repeated or continuous use of the trim motors may result in thermal cutoff and states that the motor may reset after a cooling period.[230] The Board is concerned that this additional information addressing repeated or continuous use of the trim motors may weaken or confuse the initial guidance to refrain from attempting troubleshooting measures beyond those specified in the checklist procedures.

---

[230] For additional information about the content of the Boeing flight operations bulletin, see section 1.17.2.3.1.

The Safety Board concludes that, without clearer guidance to flight crews regarding which actions are appropriate and which are inappropriate in the event of an inoperative or malfunctioning flight control system, pilots may experiment with improvised troubleshooting measures that could inadvertently worsen the condition of a controllable airplane. Accordingly, the Safety Board believes that the FAA should issue a flight standards information bulletin directing air carriers to instruct pilots that in the event of an inoperative or malfunctioning flight control system, if the airplane is controllable they should complete only the applicable checklist procedures and should not attempt any corrective actions beyond those specified. In particular, in the event of an inoperative or malfunctioning horizontal stabilizer trim system, after a final determination has been made in accordance with the applicable checklist that both the primary and alternate trim systems are inoperative, neither the primary nor the alternate trim motor should be activated, either by engaging the autopilot or using any other trim control switch or handle. Pilots should further be instructed that if checklist procedures are not effective, they should land at the nearest suitable airport. The Safety Board also believes that the FAA should direct all Certificate Management Offices (CMO) to instruct inspectors to conduct surveillance of airline dispatch and maintenance control personnel to ensure that their training and operations directives provide appropriate dispatch support to pilots who are experiencing a malfunction threatening safety of flight and instruct them to refrain from suggesting continued flight in the interest of airline flight scheduling.

2.3 Evaluation of Potential Reasons for Excessive Acme Nut Thread Wear

As previously mentioned, the condition of the recovered acme nut thread remnants indicated that approximately 90 percent of the threads had worn away before the remainder of the threads sheared off.[231] Further, calculations using the 0.033-inch end play measurement from the accident jackscrew assembly's last end play check in September 1997 and an assumed total wear of 0.133 inch at the time the worn threads were sheared off indicated that the average wear rate of the accident acme nut threads between September 1997 and the time of the accident was 0.012 inch per 1,000 flight hours. As previously discussed, the expected wear rate is about 0.001 inch per 1,000 flight hours.[232] Therefore, the Safety Board concludes that the acme nut threads on the accident airplane's horizontal stabilizer jackscrew assembly wore at an excessive rate. There is, however, no reason to believe that this excessive rate was constant throughout this time period.

In evaluating what could have caused the excessive wear of the accident acme nut threads, the Safety Board considered the potential effects of the following factors: (1) the use of Aeroshell 33 for lubrication of the jackscrew assembly, (2) acme screw thread

---

[231] For more information about the condition of the recovered acme nut threads, see section 1.16.2.6.

[232] This was the wear rate documented in 1967 test results for DC-8 and DC-9 jackscrew assemblies and was also the rate reported by McDonnell Douglas All Operators Letter (AOL) 9-2120A. For more information on these expected wear rates, see section 1.6.3.3.2.

surface finish, (3) foreign debris, (4) abnormal loading of the acme nut threads, and (5) insufficient lubrication of the jackscrew assembly.

2.3.1 Use of Aeroshell 33 for Lubrication of the Jackscrew Assembly

Task cards applicable to the lubrication of the accident jackscrew assembly for its last three scheduled lubrications (September 1999, January 1999, and June 1998) specified the use of Aeroshell 33. Findings from laboratory examinations of degraded grease found on the accident acme screw were consistent with the presence of Aeroshell 33, as well as the previously specified grease, Mobilgrease 28. Aeroshell 33 has properties that are different from those of Mobilgrease 28.[233] Therefore, Safety Board investigators considered whether Alaska Airlines' December 1997 switch to Aeroshell 33 might have caused the acme nut threads to wear faster. However, laboratory tests showed no significant adverse effects when Aeroshell 33 was used. Specifically, those tests indicated that the wear rates of test specimens lubricated with Aeroshell 33, or mixtures containing Aeroshell 33, were lower than those of test specimens lubricated with Mobilgrease 28.[234] This finding is consistent with the fact that only three of Alaska Airlines' acme nuts wore at an excessive rate after Aeroshell 33 was introduced in December 1997.[235] None of the rest of the MD-80s in Alaska Airlines' fleet showed any indication of excessive acme nut wear or any other anomalies that could be associated with the use of Aeroshell 33.

Standardized grease testing conducted by the U.S. Navy's Aerospace Materials Laboratory on mixtures of the two greases indicated that the grease mixture did not meet the industry standard for compatibility at the 90/10 and 10/90 ratios. However, test results showed that the incompatibility had an insignificant effect on the lubrication performance of the grease. These results were again confirmed by the limited number of acme nuts that wore at an excessive rate after Aeroshell 33 was introduced in September 1997. These grease mixtures would have been present on all Alaska Airlines MD-80s for some time during the switch from Mobilgrease 28 to Aeroshell 33; if Aeroshell 33 or mixtures containing Aeroshell 33 had a negative effect on acme nut wear, these effects would have been apparent throughout the fleet.

Because the recovered accident acme screw and sheared nut thread remnants had some evidence of corrosion, the surface chemistry of aluminum-bronze acme nut material was evaluated. Tests conducted by the U.S. Navy's Aerospace Materials Laboratory and Science Applications International Corporation determined that Aeroshell 33 was not

---

[233] Boeing's MD-80 Maintenance Manual specified the use of lubricants that meet the specifications of MIL-G-81322. Mobilgrease 28 meets these specifications, whereas Aeroshell 33 meets the specifications of MIL-G-23827.

[234] For more information about these tests, see section 1.16.8.2.

[235] Only three Alaska Airlines jackscrew assemblies are known to have worn at a higher-than-expected rate: the accident assembly and the assemblies from airplanes N981AS and N982AS. For more information about the acme nuts from airplanes N981AS and N982AS, see section 1.18.4.

corrosive to aluminum bronze.[236] It is noteworthy that none of the Alaska Airlines acme screws or nuts examined by the Safety Board that had been lubricated with Aeroshell 33, including those from two other Alaska Airlines airplanes that wore at an excessive rate, revealed evidence of corrosion. However, immersion of unlubricated metal surfaces in salt water is known to cause corrosion. Therefore, the corrosion on the acme screw and nut thread remnants was very likely caused by exposure to seawater during the 7 1/2 days before they were recovered, and the use of Aeroshell 33 to lubricate the jackscrew assembly was not a factor in the excessive wear of the accident acme nut.

Nonetheless, because of the potential for improper lubrication practices and inappropriate grease selection to have adverse safety effects, the Safety Board continues to be concerned that Alaska Airlines began using Aeroshell 33, and the FAA did not object to its use, without sufficient research, testing, or followup tracking and evaluation of its in-service performance to demonstrate its acceptability.[237] This lack of research, testing, and followup in connection with the December 1997 grease change is all the more disconcerting considering that, in April 1996, Alaska Airlines had significantly extended the lubrication interval for the horizontal stabilizer and associated components from 1,600 flight hours to 8 months (the equivalent of about 2,250 flight hours). That interval extension was presumably based on the conclusion that the previous long-term performance of the grease then in use (Mobilgrease 28) demonstrated that it would remain effective in those areas for the extended interval. However, Alaska Airlines had no previous experience with Aeroshell 33 in those areas and, therefore, could not have known whether this extended interval would also be appropriate for the performance of Aeroshell 33.

2.3.2 Acme Screw Thread Surface Finish

Because of previous instances of excessive acme nut thread wear on MD-11 jackscrew assemblies resulting from improper acme screw thread surface finish,[238] Safety Board investigators considered this as a possible factor in the excessive wear of the accident nut threads. However, a detailed metallurgical examination of the accident acme screw found no evidence that the thread surface finish was out of production specifications.

2.3.3 Foreign Debris

Because the excessive wear of two Hawaiian Airlines DC-9 acme nuts was apparently associated with the presence of abrasive grit-blasting material in the jackscrew assembly,[239] Safety Board investigators considered whether the abrasive effects of foreign

---

[236] For more information about these tests, see section 1.16.8.2.

[237] For more information about the circumstances regarding and the procedures followed for this grease change, see section 1.6.3.2.2.

[238] For more information about these instances of premature wear on MD-11 jackscrew assemblies, see section 1.18.5.

[239] For more information about these two acme nuts, see section 1.16.6.1.

debris might have been a factor in the excessive wear of the accident acme nut threads. However, examinations and tests of grease samples taken from the accident acme screw found no evidence of foreign debris that would have abraded the acme nut threads during operation. Metallic particles found in grease residue from the acme nut's grease fitting counterbore were from the nut itself and were consistent with sliding wear, not abrasive wear. Further, although the Hawaiian Airlines jackscrew assemblies showed evidence of abrasive wear on the acme screw threads and the underside of the acme nut threads, which normally do not wear, there was no such evidence on the accident acme screw and nut threads.

Although the abrasive effects of foreign debris was not a factor in the excessive wear of the accident jackscrew assembly, the Safety Board remains concerned that this could play a role in the excessive wear of other jackscrew assemblies. This issue, and the related issue of the adequacy of the current end play check interval, are further discussed in section 2.4.2.

2.3.4 Abnormal Loading of the Acme Nut Threads

Safety Board investigators also considered the possibility that abnormal loading on the jackscrew assembly may have caused the acme nut threads to wear excessively. This scenario was evaluated following reports from Alaska Airlines of a "wobbling" acme screw.[240] The wobble appeared as a slight rocking motion of the acme nut as the screw was rotated within it. The investigators determined that the wobble in that and other jackscrew assemblies, including the accident jackscrew assembly, was associated with a torque tube wear band. It was hypothesized that this wobble might alter the engagement of the acme nut and screw threads, thus accelerating thread wear. However, analysis of the jackscrew assembly revealed that the gimbal rings, which hold the acme nut in place, accommodate any such wobble and prevent wobbling from resulting in any abnormal loading. This was confirmed when comparing wear rates of acme nuts from assemblies that had torque tubes with little or no wear band and those with deeper wear bands. The comparison showed no statistically significant difference in corresponding average wear rates.

2.3.5 Summary of Possibilities Ruled Out as Reasons for the Excessive Acme Nut Thread Wear

As a result of the findings discussed in sections 2.3.1 through 2.3.4, the Safety Board concludes that Alaska Airlines' use of Aeroshell 33 for lubrication of the jackscrew assembly, acme screw thread surface finish, foreign debris, and abnormal loading of the acme nut threads were not factors in the excessive wear of the accident acme nut threads.

---

[240] For more information about the wobbling acme screw, see section 1.16.6.2.

2.3.6 Insufficient Lubrication of the Jackscrew Assembly

It is generally understood that a lack of lubrication of moving parts significantly increases wear. Laboratory tests confirmed that the unlubricated wear rate of aluminum-bronze test specimens, under pressure conditions associated with the normal flight loads on the jackscrew assembly, was roughly 10 times higher than the wear rate of test specimens lubricated with any tested grease type or combination of greases.[241] As discussed in section 1.6.3.3.2, previous instances of excessive and accelerated wear of DC-9 and MD-80 acme nuts has been attributed to, or associated with concerns about, insufficient lubrication. Thus, the accident acme nut threads' excessive and accelerated wear is consistent with a lack of sufficient lubrication. Furthermore, as discussed below, there were additional indications that the accident jackscrew assembly had not been adequately lubricated before the accident.

First, no evidence of any type of grease in any condition, either semi-fluid (that is, fresh) or dry/solid (that is, old/degraded), was found inside the acme nut or on the working region of the recovered acme screw.[242] Further, in photographs taken immediately after recovery and during initial inspection of the jackscrew assembly, no grease is visible on the working areas of the acme screw. (See figures 12a and 12b.) Detailed metallurgical examination revealed only small flakes of dried and hardened grease attached to some of the acme nut thread remnants wrapped around the acme screw, but no evidence of grease capable of providing significant lubrication.

These findings contradict the assertions of some Alaska Airlines representatives who stated that they observed or felt a greaselike substance on the surface of the acme screw immediately after it was recovered. However, the Safety Board notes that seawater, fuel, or hydraulic fluid from the accident wreckage could have caused the acme screw to appear shiny and that this sheen might have been mistaken for grease. To preserve the condition of the acme screw and thread remnants, they were rinsed with fresh water immediately after recovery, but that rinsing was accomplished by directing the water from a hose similar to a garden hose over the gearbox and allowing the water to flow down over the acme screw; the stream of water used to rinse the screw was not directly sprayed on it. The acme screw may have been rinsed again with a garden-type hose after it was brought on shore, but it was never subjected to any type of high-pressure wash. It should be noted that photographs of the acme screw taken after these rinses show sand still adhered to the surfaces. (See figure 14.)

Although the jackscrew assembly had been immersed in seawater for 7 1/2 days before it was recovered, laboratory tests at the U.S. Navy's Aerospace Materials Laboratory showed that immersion in seawater does not significantly alter or remove grease.[243] This was demonstrated by the fact that semi-fluid grease was found adhered to

---

[241] For more information about these tests, see section 1.16.8.3.

[242] As previously stated, the acme screw working region is that part of the screw that can come in contact with the acme nut during its operation between the upper and lower mechanical stop limits.

[243] For more information about these tests, see section 1.16.8.1.

airplane components on the accident airplane even after they had been immersed in seawater for 7 1/2 days. Specifically, some evidence of semi-fluid grease was found outside the working region on the extreme lower end of the acme screw, and significant amounts of intact semi-fluid grease were visible on other airplane components. The Safety Board notes that a coating of grease was still clearly visible on an acme screw recovered from a China Northern MD-80 airplane that crashed on May 7, 2002, into a bay near Dalian, China.[244] The acme screw from that airplane had been immersed in seawater for 5 days.

Second, the recovered acme nut's grease fitting passageway and counterbore contained a dried, black, claylike substance consistent with degraded grease. If the acme nut's grease fitting had recently been used to lubricate the jackscrew assembly, as it should have been during the jackscrew assembly's last lubrication on September 14, 1999 (as well as during all previous lubrications), the counterbore should have contained fresh grease. The contents of the counterbore area should not have been affected by impact or salvage damage because the area was protected externally by the grease fitting on one side and was isolated internally from direct contact on the bore side.

Third, postaccident interviews with the SFO mechanic who was responsible for performing the last lubrication of the accident airplane's horizontal stabilizer components, including the jackscrew assembly, revealed his lack of knowledge about how to properly perform the procedure. For example, he indicated that the lubrication task took about 1 hour, whereas Boeing documents and testimony indicated that, when properly done, the task should take more than 4 person hours. The SFO mechanic also stated that he did not recall checking to see if grease was coming out of the top of the acme nut, as specified in the general lubrication guidance. Further, in the recovered wreckage, about 40 percent of the other grease fittings on the elevators and horizontal stabilizer included on the same task card as the acme screw and nut contained either dry or semi-dry grease. In comparison, grease fittings on the rudder that had been lubricated in January 1999 by a different mechanic contained mostly fresh or semi-fresh grease. This suggests that the SFO mechanic who was responsible for lubricating the jackscrew assembly in September 1999 did not adequately perform the task. It is noteworthy that the other two Alaska Airlines jackscrew assemblies with higher-than-expected wear rates had received their last horizontal stabilizer lubrication servicing at OAK during C checks, but both had received the lubrication before that at SFO. The lubrication of one of those two assemblies at SFO was accomplished by the same mechanic who was responsible for the last lubrication of the accident airplane's jackscrew assembly.

On the basis of all of the factors discussed in this section, the Safety Board concludes that there was no effective lubrication on the acme screw and nut interface at the time of the Alaska Airlines flight 261 accident. Further, the Board notes that the inadequate lubrication of the accident jackscrew assembly was not an isolated occurrence. The recent discovery by Alaska Airlines of inadequately lubricated MD-80 rudder trim tab

---

[244] The last scheduled lubrication of the jackscrew assembly on the China Northern airplane was during an A check that took place sometime between March 27 and April 2, 2002, and the specified grease was Aeroshell 5 (MIL-G-3545C).

hinge support bearings[245] demonstrates that deficiencies continue to exist in Alaska Airlines' lubrication practices.

2.3.6.1 Analysis of How Many Recently Scheduled Lubrications Might Have Been Missed or Inadequately Performed Before the Accident

The service history of Alaska Airlines' MD-80 fleet indicates that jackscrew assembly lubrications remained effective for the 8-month lubrication interval (equivalent to approximately 2,550 flight hours) in effect before the accident. Otherwise, Alaska Airlines' entire MD-80 fleet would have experienced excessive acme nut wear, which was not the case. At the time of the accident, only 4 months had elapsed since the accident airplane's last scheduled jackscrew assembly lubrication in September 1999. The condition of the recovered jackscrew assembly, specifically, the absence of any grease inside the acme nut or on the working areas of the acme screw, and the degraded grease inside the acme nut's grease fitting passageway and counterbore, indicates that at least that lubrication was missed or inadequate.

Then number of lubrication opportunities beyond the one in September 1999 that might have been performed inadequately or missed entirely is less certain.[246] However, it is apparent that the abnormally high wear of the accident acme nut threads likely occurred over a period of time that extended beyond the 4 months between the September 1999 lubrication and the accident date. Laboratory tests indicate that the wear rate for unlubricated parts[247] cannot account for the amount of wear exhibited by the accident acme nut thread remnants if it had occurred only since the last scheduled lubrication. This indicates that unlubricated wear rates were likely experienced over a longer period of time and, therefore, more than just the last scheduled lubrication was missed or inadequately performed. The Safety Board could not determine exactly how many more scheduled lubrications were missed or inadequately performed. However, the Safety Board concludes that the excessive and accelerated wear of the accident jackscrew assembly acme nut threads was the result of insufficient lubrication, which was directly causal to the Alaska Airlines flight 261 accident.

2.3.6.2 Alaska Airlines' Lubrication Interval Extension

DC-9 test results and DC-8 service history indicated that frequent lubrication of the jackscrew assembly would allow the acme screw to meet its original design life of 30,000 flight hours. DC-9 certification documents, including Douglas Process Standard 3.17-49 (issued August 1, 1964), specified a lubrication interval for the

---

[245] For more information, see section 1.6.3.2.3.

[246] Because of the absence of any end play measurements since the October 1997 measurement, there is little upon which to base a wear rate estimate for periods of time beginning after the October 1997 measurement. Further uncertainty is introduced by the amount of wear exhibited by the thread remnants (approximately 90 percent of their original thickness had worn). The Safety Board's FEA indicated that the load distribution, and thus the associated acme nut thread wear rates, changes dramatically after a wear of approximately 0.095 inch (approximately 74 percent worn), thereby invalidating the traditionally assumed relationship between end play and wear rate.

[247] For more information about these tests, see section 1.16.8.3.

jackscrew assembly of 300 to 350 flight hours. However, the 300- to 350-flight-hour recommended lubrication interval for the DC-9 was not contained in the manufacturer's initial on-aircraft maintenance planning (OAMP) documents for the DC-9 or the MD-80. Instead, those documents specified a lubrication interval of 600 to 900 flight hours.

As discussed in section 1.6.3.2.1.2, in 1987, Alaska Airlines' lubrication interval for horizontal stabilizer components, including the jackscrew assembly, was every 500 flight hours, consistent with the manufacturer's recommendation in the Maintenance Review Board (MRB) report and OAMP documents derived from Maintenance Steering Group (MSG)-2 guidance, which recommended a lubrication interval of 600 to 900 flight hours. In 1988, Alaska Airlines' lubrication interval increased to every 1,000 flight hours (a 100 percent increase); in 1991, to every 1,200 flight hours (an additional 20 percent increase); and, in 1994, to every 1,600 flight hours (an additional 33 percent increase). In 1996, the interval was changed to 8 months with no specified flight-hour limit. Based on fleet utilization at the time, 8 calendar months equated to about 2,550 flight hours, an additional increase of greater than 59 percent. Thus, at the time of the accident, Alaska Airlines' lubrication interval for the jackscrew assembly was more than 400 percent greater than it was in 1987.

The investigation did not determine what type of information, if any, was presented as justification for the lubrication interval extensions in 1988, 1991, and 1994. However, according to the FAA principal maintenance inspector (PMI) for Alaska Airlines, who reviewed and accepted the 1996 interval extension, Alaska Airlines presented documentation of the manufacturer's recently extended recommended interval as justification for its increase. (The extended recommended lubrication interval in the MRB report and OAMP document derived from the MSG-3 guidance was every C check, or every 3,600 flight hours.)

Testimony at the Safety Board's public hearing and Boeing documents indicated that the original design engineers' recommended lubrication interval was not considered during the MRB-3 decision-making process regarding the extension of the manufacturer's recommended interval. Further, Boeing design engineers were not consulted about nor aware of the extended 3,600-flight-hour MRB-3 recommended lubrication interval. Although Alaska Airlines' extended lubrication and end play check intervals have now been superseded by the 650-flight-hour lubrication and 2,000-flight-hour end play check intervals specified in Airworthiness Directive (AD) 2000-15-15, the Safety Board is concerned that there is no mechanism in place to prevent similar unsafe interval extensions for other maintenance tasks. This issue and associated safety recommendations are discussed further in section 2.4.1.

2.3.6.2.1 Safety Implications of Lubrication Interval Extension

As grease is used in a system, it loses its effectiveness over time and requires replacement. Longer lubrication intervals increase the likelihood that a missed or inadequate lubrication will result in excessive wear. Conversely, shorter lubrication intervals increase the likelihood that, even if a lubrication is missed or inadequately performed, the existing grease will remain effective until the next scheduled lubrication.

The Safety Board notes that at the time of Alaska Airlines' increase to an 8-month lubrication interval, it was the only U.S. airline that had a calendar-time lubrication interval with no accompanying flight-hour limit and no specification, "whichever comes first." A calendar-time lubrication interval can degrade the margin of safety because wear is directly related to aircraft usage, or flight hours, and not to calendar time. Also, a purely calendar-based interval does not account for increases in flight hours that result from increased airplane utilization. Conversely, intervals based on flight hours, or on calendar time with an accompanying flight-hour time limit and the proviso, "whichever comes first," ensure that the flight-hour limit will not be exceeded as a result of increased airplane utilization. Thus, unless a maximum utilization is also specified, calendar-time intervals are inappropriate for certain tasks, such as lubrication or inspections for fatigue, when component deterioration is related to usage, not time.[248]

In sum, at the time of the accident, Boeing's recommended lubrication interval for the MD-80 jackscrew assembly was every 3,600 flight hours, about 4 to 6 times longer than Douglas' original recommendation in the MSG-2 OAMP document of every 600 to 900 flight hours. Alaska Airlines' lubrication interval for the MD-80 jackscrew assembly—although it was still less than the manufacturer's recommended interval in the MSG-3 OAMP document of 3,600 flight hours—was about 3 to 4 times longer than Douglas' originally recommended lubrication interval, resulting in a significant decrease in the MD-80 fleet's ability to tolerate missed or inadequate lubrications. The Board notes that the negative safety implications of the ongoing lubrication interval extensions were magnified by the simultaneous ongoing extensions of the end play check interval in that there would be fewer and fewer opportunities to discover and address any excessive wear resulting from lubrication deficiencies.

Therefore, the Safety Board concludes that Alaska Airlines' extensions of its lubrication interval for its MD-80 horizontal stabilizer components and the FAA's approval of these extensions, the last of which was based on Boeing's extension of the recommended lubrication interval, increased the likelihood that a missed or inadequate lubrication would result in excessive wear of jackscrew assembly acme nut threads and, therefore, was a direct cause of the excessive wear and contributed to the Alaska Airlines flight 261 accident.

2.3.6.3 Adequacy of Lubrication Procedures

The horizontal stabilizer lubrication procedure specifies that, after the access doors are opened, grease is to be applied to the acme nut grease fitting under pressure. The procedure further specifies the brush application of a light coat of grease to the acme screw threads and operation of the trim system through its full range of travel to distribute the grease over the length of the acme screw.

---

[248] In contrast, some maintenance tasks are intended to prevent or control conditions (such as corrosion or deterioration based on aging) that are based solely on the passage of time. Such tasks might appropriately be tied to calendar-time intervals without any flight-hour limits.

Safety Board investigators observed maintenance personnel from two MD-80 operators perform jackscrew assembly lubrications and discussed the lubrication procedure with those and other maintenance personnel from those operators. Investigators noted many differences in the methods used by the personnel to accomplish certain steps in the lubrication procedure, including the manner in which grease was applied to the acme nut fitting and screw and the number of times the trim system was cycled to distribute the grease. Several of the methods observed by or reported to investigators did not involve application of grease to the entire length of the acme screw and cycling the trim system several times.

Laboratory demonstrations designed to compare the effectiveness of various methods of lubricating the jackscrew assembly found that application of a complete coating of grease over all exposed threads, filling the thread valleys, followed by the cycling of the trim system several times, maximized the distribution of the grease over the length of the acme screw. In contrast, the observations and demonstrations established that applying grease only through the acme nut grease fitting and then cycling the trim system several times did not distribute an adequate amount of grease over the remainder of the acme screw. Although both methods of grease application are specified in the lubrication procedure, if a mechanic mistakenly believed that lubricating only through the acme nut grease fitting was adequate, the acme screw and nut would receive insufficient lubrication.

The extent to which these deficiencies in the lubrication procedure may have played a role in the inadequate lubrication of the accident jackscrew assembly could not be determined. However, in an October 1, 2001, safety recommendation letter, the Safety Board expressed its concern that the current lubrication procedure was not adequate to ensure consistent and thorough lubrication of the jackscrew assembly by all operators and issued Safety Recommendation A-01-41, which asked the FAA to require Boeing to "revise the lubrication procedure for the horizontal stabilizer trim system of Douglas DC-9, McDonnell Douglas MD-80/90, and Boeing 717 series airplanes to minimize the probability of inadequate lubrication." In a December 12, 2001, letter, the FAA responded that it agreed with the intent of Safety Recommendation A-01-41 and that it was "working with the Boeing Commercial Airplane Group to rewrite the lubrication procedures to the optimal standard." On June 14, 2002, the Safety Board classified the recommendation "Open—Acceptable Response."

After the issuance of Safety Recommendation A-01-41, Safety Board investigators continued to evaluate the adequacy of the current lubrication procedure and identified three additional areas that should be addressed.

First, the Safety Board is concerned about the wear debris produced as a result of the wear process and other foreign debris that may accumulate in the grease over time. Because material that infiltrates the working grease reduces its lubricating effectiveness, it is desirable to flush out these materials before fresh grease is added (that is, to completely replace used, less effective grease with fresh, more effective grease).[249] The Board notes that the current jackscrew lubrication procedure does not stipulate the removal of used grease from the acme screw before the application of fresh grease.[250] The Board is aware that Boeing is developing a revised lubrication procedure for the jackscrew assembly that includes removal of the used grease before application of the fresh grease. Although the revised procedure, if properly performed, should improve the effectiveness of the lubrication, it is not yet known whether the FAA will require the use of this improved procedure.

The Safety Board concludes that when lubricating the jackscrew assembly, removal of degraded grease from the acme screw before application of fresh grease will increase the effectiveness of the lubrication. Therefore, the Safety Board believes that as part of the response to Safety Recommendation A-01-41, the FAA should require operators of DC-9, MD-80/90, and 717 series airplanes to remove used grease from the jackscrew assembly acme screw and flush degraded grease and particulates from the acme nut before applying fresh grease.

Second, Safety Board investigators noted that when they attempted to perform the lubrication procedure, it was difficult to insert a hand through the access panel openings because of their size.[251] They further noted that after a hand was inserted, it blocked the view of the jackscrew assembly, requiring the task to be accomplished primarily by "feel." The Board is aware that Boeing is developing a modification for an expanded access panel. The Safety Board concludes that a larger access panel would facilitate the proper accomplishment of the jackscrew assembly lubrication task. Therefore, the Safety Board believes that, as part of the response to Safety Recommendation A-01-41, the FAA should require operators of DC-9, MD-80/90, and 717 series airplanes, in coordination with Boeing, to increase the size of the access panels that are used to accomplish the jackscrew assembly lubrication procedure.

Third, this investigation has highlighted the need for improved methods for ensuring that jackscrew assembly lubrications are accomplished properly at scheduled lubrication intervals. Currently, the lubrication procedure is generally performed and signed off by a single maintenance technician, and that technician's work is not required to be inspected.

Although the Safety Board cannot be certain that such a requirement would have prevented the Alaska Airlines flight 261 accident, the Safety Board concludes that if the jackscrew assembly lubrication procedure were a required inspection item for which an inspector's signoff is needed, the potential for unperformed or improperly performed lubrications would be reduced. Therefore, because of the critical importance of adequately lubricating the jackscrew assembly and the potentially catastrophic effects of excessive acme nut wear resulting from insufficient lubrication, the Safety Board believes that the

---

[249] This finding was further supported by interviews with representatives of the largest manufacturer of nonaviation jackscrew assemblies, Nook Industries. Nook representatives told Safety Board investigators that purging old grease out of the jackscrew assembly, and then refreshing the acme screw with new grease, increases the life of the acme nut.

[250] The current jackscrew lubrication procedure does call for the application of new grease into the nut, via the grease fitting, until grease is observed extruding from the nut; however, this does not remove the old grease from the acme screw surface or a large portion of the nut.

[251] Two access panel openings (one 4 inches by 6 inches and the other 8 inches by 7 inches) are generally used to accomplish the lubrication procedure.

FAA should establish the jackscrew assembly lubrication procedure as a required inspection item that must have an inspector's signoff before the task can be considered complete.

2.4 Monitoring Acme Nut Thread Wear

Because the MD-80 jackscrew assembly's structural function is critical to the safety of flight, and that structural function cannot be maintained without proper acme nut and screw thread engagement, it is essential that acme nut thread wear be regularly monitored. The failure to adequately monitor acme nut thread wear may result in continued operation of an airplane with excessive nut thread wear. As demonstrated by the Alaska Airlines flight 261 accident, because no other structure performs the function of the jackscrew assembly, the loss of acme nut and screw engagement as a result of excessive wear will most likely have catastrophic results.

As previously discussed, in-service acme nut thread wear is monitored by performing an on-wing end play check procedure at specified intervals. Before 1967, there were no required periodic inspections. The jackscrew assembly had an expected service life of 30,000 hours, at which time the end play measurement was expected to be 0.0265 inch. However, as a result of higher-than-expected wear rates reported in 1966, in 1967, Douglas developed the end play check procedure and increased the maximum permissible end play measurement at which a jackscrew assembly could remain in service to 0.040 inch. (The minimum end play measurement remained at 0.003 inch.)[252] In AOL 9-48A, Douglas stated that jackscrew assemblies could remain in service as long as the end play measurement remained within these tolerances.

2.4.1 Alaska Airlines' Preaccident End Play Check Intervals

Alaska Airlines has consistently required end play checks at every other C check. However, the length of its C-check interval has changed over time. As a result of those C-check interval changes, between 1985 and 1996 Alaska Airlines increased its end play check interval by almost 200 percent.

In 1985, when Alaska Airlines' MD-80 maintenance program was initially approved by the FAA, C checks were conducted every 2,500 flight hours; therefore, end play checks were performed every 5,000 flight hours. By July 1988, C-check intervals had been extended to 13 months, with no corresponding flight-time limit; therefore, end play checks were performed every 26 months, which was approximately 6,400 flight hours, based on the airplane utilization rate at that time. Although FAA approval would have been required for this C-check interval extension, no information was available regarding what documentation, if any, Alaska Airlines presented to the FAA to justify the extension.

---

[252] As discussed in section 1.6.3.3.2, as a result of these higher-than-expected wear rates, Douglas also changed some of the materials specifications and manufacturing processes for the acme screw (increased heat treating and nitriding) to reduce wear.

In 1996, Alaska Airlines' C-check interval was extended to 15 months; therefore, end play checks were only required every 30 months. Because Alaska Airlines' airplane utilization had also increased, this 30-month interval was equivalent to approximately 9,550 flight hours. Alaska Airlines sought and obtained advance FAA approval for this C-check interval extension. Alaska Airlines' director of reliability and maintenance programs testified at the public hearing that Alaska Airlines had presented the FAA with a data analysis package based on the maintenance histories of five sample airplanes to justify the C-check interval extension. He indicated that individual maintenance tasks tied to C-check intervals (such as the end play check) were not separately considered in connection with the extension.[253] Thus, the FAA's approval of the 1996 C-check extension also effectively constituted approval to extend the end play check from 26 to 30 months, a 15 percent increase, and, more importantly, a 55 percent increase in flight-hour intervals, from approximately 6,400 to 9,550 flight hours.

Although Alaska Airlines' extended 30-month end play check interval was consistent with the manufacturer's recommended calendar-time limit, the resulting 9,550-flight-hour interval far exceeded the manufacturer's recommended flight-hour interval, which at that time was 7,000 or 7,200 flight hours (depending on whether MSG-2 or MSG-3 guidance was used). At that time, Alaska Airlines had the second highest end play check interval of all operators of DC-9, MD-80/90, and 717 series airplanes. Further, Alaska Airlines was the only U.S. carrier at that time that did not have an accompanying flight-hour limit with the specification, "whichever comes first," for its end play check interval. As discussed previously in connection with Alaska Airlines' lubrication interval, use of only a calendar-time interval does not account for increased airplane utilization rates and could result in a lower safety margin than intended.

The Safety Board notes that, because of Alaska Airlines' extended end play check interval of 30 months, or about 9,550 flight hours, after the accident airplane's last end play check in September 1997, the jackscrew assembly would not have been required to undergo another end play check until March 2000. Between the time that the 0.040- and 0.033-inch end play measurements were recorded in September 1997 and the time of the accident, the airplane had been flown for 28 months, nearly 9,000 flight hours. During this time, the acme nut thread wear progressed to failure.

In light of what has been learned in this investigation, it is now apparent that the manufacturer's previously recommended end play check intervals of 7,000 or 7,200 flight hours were not adequate.[254] Nonetheless, the Safety Board notes that if Alaska Airlines had not extended its end play check interval to beyond the recommended interval, the airplane would have been required to undergo an end play check at least 1,800 to 2,000 flight hours before the accident, and the excessive end play could have been identified at

---

[253] However, based on the history of maintenance discrepancies noted for the five sample airplanes, two maintenance tasks were identified as inappropriate for extension, and these tasks were converted to stand-alone items to be performed at shorter intervals. The two maintenance tasks identified as requiring shorter intervals based on service history were (1) lubrication of the bent-up, trailing-edge wing doors and (2) lubrication of the bearings and bushings in the elevator hinges.

[254] This issue is discussed further in the next section.

that time. Thus, the Safety Board concludes that Alaska Airlines' extension of the end play check interval and the FAA's approval of that extension allowed the accident acme nut threads to wear to failure without the opportunity for detection and, therefore, was a direct cause of the excessive wear and contributed to the Alaska Airlines flight 261 accident.

2.4.1.1 Adequacy of Existing Process for Establishing Maintenance Task Intervals

The Safety Board is concerned that the absence of any significant maintenance history pertaining to the jackscrew assembly was apparently considered by Alaska Airlines and the FAA as sufficient justification to extend the end play check interval as part of the C-check interval extension. In general, the absence of maintenance history should not be considered adequate justification to extend the interval for the performance of a critical maintenance task. Any significant maintenance change associated with a critical flight control system should be independently analyzed and supported by technical data demonstrating that the proposed change will not present a potential hazard. Therefore, any maintenance task change related to the jackscrew assembly, which is an essential element of a critical flight control system, should be handled in this manner.

The Safety Board is further concerned that the MSG and MRB-based process by which manufacturers develop initial and revised recommended maintenance task intervals[255] resulted in significant extensions of both the lubrication and end play check intervals without any such analysis or support. As discussed in section 1.6.3.2.1, testimony at the Safety Board's public hearing and Boeing documents indicated that Douglas' original recommended lubrication interval of 600 to 900 flight hours was not considered during the MSG-3 decision-making process to extend the recommended interval to 3,600 hours. Further, Boeing design engineers were not consulted about nor aware of the escalated lubrication interval specified in the MSG-3 documents. The FAA's MD-80 MRB chairman testified at the public hearing that the escalation of C-check intervals in the MSG-3 MD-80 MRB did not involve a task-by-task analysis of each task (such as the jackscrew lubrication task and end play check) that would be affected by the changed interval.

The Safety Board concludes that Alaska Airlines' end play check interval extension should have been, but was not, supported by adequate technical data to demonstrate that the extension would not present a potential hazard. The Safety Board further concludes that the existing process by which manufacturers revise recommended maintenance task intervals and by which airlines establish and revise these intervals does not include task-by-task engineering analysis and justification and, therefore, allows for the possibility of inappropriate interval extensions for potentially critical maintenance tasks. In addition, the Board notes that the FAA plays a limited role in this process compared to the role it plays in the initial certification process.

---

[255] For more information about this process, see section 1.6.3.1.1.

Therefore, the Safety Board believes that the FAA should review all existing maintenance intervals for tasks that could affect critical aircraft components and identify those that have been extended without adequate engineering justification in the form of technical data and analysis demonstrating that the extended interval will not present any increased risk and require modification of those intervals to ensure that they (1) take into account assumptions made by the original designers, (2) are supported by adequate technical data and analysis, and (3) include an appropriate safety margin that takes into account the possibility of missed or inadequate accomplishment of the maintenance task. In conducting this review, the FAA should also consider original intervals recommended or established for new aircraft models that are derivatives of earlier models and, if the aircraft component and the task are substantially the same and the recommended interval for the new model is greater than that recommended for the earlier model, treat such original intervals for the derivative model as "extended" intervals. The Safety Board further believes that the FAA should conduct a systemic industrywide evaluation and issue a report on the process by which manufacturers recommend and airlines establish and revise maintenance task intervals and make changes to the process to ensure that, in the future, intervals for each task (1) take into account assumptions made by the original designers, (2) are supported by adequate technical data and analysis, and (3) include an appropriate safety margin that takes into account the possibility of missed or inadequate accomplishment of the maintenance task.

The Safety Board also believes that the FAA should require operators to supply the FAA, before the implementation of any changes in maintenance task intervals that could affect critical aircraft components, technical data and analysis for each task demonstrating that none of the proposed changes will present any potential hazards, and obtain written approval of the proposed changes from the PMI and written concurrence from the appropriate FAA Aircraft Certification Office.

2.4.2 Adequacy of Current End Play Check Intervals

The Safety Board recognizes that the FAA acted promptly following the Alaska Airlines flight 261 accident by issuing ADs that shortened the end play check interval to 2,000 flight hours. However, evidence collected during this investigation suggests that acme nut thread wear at a higher-than-expected rate could allow a potentially dangerous level of wear to occur in less than 2,000 hours.

As previously discussed, two jackscrew assemblies installed sequentially on the same Hawaiian Airlines DC-9 airplane that were removed because of high end play measurements were found to have worn at an unprecedented rate. The first acme nut had an approximate wear rate of 0.015 inch per 1,000 flight hours, and the second had an approximate wear rate of 0.008 inch per 1,000 flight hours.[256] These wear rates are about 15 and 8 times greater, respectively, than the expected wear rate of about 0.001 inch per 1,000 flight hours. This accelerated wear was attributed to the presence of grit-blasting

---

[256] For more information about these jackscrew assemblies, see section 1.16.6.1.

material that had been introduced inadvertently into the jackscrew assembly and become embedded in the grease on the acme screw.

If the first excessively worn jackscrew assembly had measured just within, rather than beyond, the 0.040-inch limit at the last end play check, and assuming the acme nut threads had continued to wear at an approximate rate of 0.015 inch per 1,000 flight hours, the end play measurement would have been about 0.069 inch by the time of its next scheduled end play check 2,000 flight hours later. Further, assuming that this end play check was either missed or improperly accomplished and the jackscrew assembly remained in service despite an end play of 0.069 inch and continued to wear at the same rate, the end play measurement would have been about 0.099 inch by the time of the next scheduled end play check. Moreover, there is no basis for assuming that the wear rate of this Hawaiian Airlines jackscrew assembly represents the maximum possible acme nut thread wear rate. Just as the wear rate and wear mechanism on the Hawaiian Airlines jackscrew assembly was unprecedented and unanticipated, there may be other unprecedented and unanticipated wear rates and mechanisms that could also result in excessive or accelerated wear. Therefore, it is possible that acme nut threads could wear at an even faster rate than the Hawaiian Airlines acme nut threads, possibly even to catastrophic limits[257] in 2,000 flight hours.

To establish an appropriately conservative end play check interval, the uncertainties regarding possible wear mechanisms and the maximum possible wear rate must be considered in addition to the significant possibility of inaccurate end play measurements. The Safety Board notes that, when failure mechanisms are known and clearly defined, standard industry practice to ensure a damage tolerant design is to have a safety margin that allows for two complete inspection cycles before the predicted failure time.[258] Thus, even if one inspection is missed or inadequately performed, there will be at least one other opportunity to detect and correct the condition. However, when uncertainties exist in the failure mechanism, as in the case of acme nut thread wear, standard industry practice is to increase the safety margin to account for the reduced level of confidence in the predicted failure time.

In a June 26, 2001, letter to the Safety Board, the FAA stated that it believed the 2,000-flight-hour interval provided an acceptable level of safety, citing the "robust design" of the acme nut and the fact that it could safely carry normal flight loads even when worn beyond 0.080 inch. The FAA also stated that the 650-flight-hour inspection and

---

[257] According to the Safety Board's study of thread stress and deformation, the acme nut threads will begin to deflect and begin the process of sliding over the acme screw threads at a wear level of about 0.093 inch.

[258] According to the FAA's *Damage Tolerance Assessment Handbook, Volume I,* issued in February 1999, "damage tolerance refers to the ability of the design to prevent structural cracks from precipitating catastrophic fracture when the airframe is subjected to flight or ground loads. Transport category airframe structure is generally made damage tolerant by means of redundant ('fail safe') designs for which the inspection intervals are set to provide at least two inspection opportunities per number of flights or flight hours it would take for a visually detectable crack to grow large enough to cause a failure in flight." Although this refers to cracks, the Safety Board notes that the damage tolerance principles can be applied to acme nut wear.

lubrication interval in AD 2000-15-15 provided frequent opportunities (in addition to the end play check every 2,000 flight hours) to detect wear debris and, therefore, possible excessive wear. Nonetheless, the Safety Board is concerned that significantly higher-than-expected wear—wear even greater than that of the Hawaiian Airlines jackscrew assemblies—could result from foreign-object contamination, such as grit blast, or from other factors that have not yet been identified.

The Safety Board concludes that, because of the possibility that higher-than-expected wear could cause excessive wear in less than 2,000 flight hours and the additional possibility that an end play check could be not performed or improperly performed, the current 2,000-flight-hour end play check interval specified in AD 2000-15-15 may be inadequate to ensure the safety of the DC-9, MD-80/90, and 717 fleet. Therefore, the Safety Board believes that, pending the incorporation of a fail-safe mechanism in the design of the DC-9, MD-80/90, and 717 horizontal stabilizer jackscrew assembly, as recommended in Safety Recommendation A-02-49 in this report, the FAA should establish an end play check interval that (1) accounts for the possibility of higher-than-expected wear rates and measurement error in estimating acme nut thread wear and (2) provides for at least two opportunities to detect excessive wear before a potentially catastrophic wear condition becomes possible.

To establish an appropriate end play check interval, it is necessary to monitor end play measurements over time to identify any excessive or unanticipated wear rates and to continue evaluating the reliability and validity of end play measurements. Therefore, the FAA, Boeing, maintenance facilities that overhaul jackscrew assemblies, and operators should closely evaluate the measurement data currently being reported pursuant to AD 2000-15-15. The Safety Board concludes that the continued collection and analysis of end play data are critical to monitoring acme nut thread wear and identifying excessive or unexpected wear rates, trends, or anomalies. Therefore, the Safety Board believes that the FAA should require operators to permanently (1) track end play measurements according to airplane registration number and jackscrew assembly serial number, (2) calculate and record average wear rates for each airplane based on end play measurements and flight times, and (3) develop and implement a program to analyze these data to identify and determine the cause of excessive or unexpected wear rates, trends, or anomalies. The Safety Board further believes that the FAA should require operators to report this information to the FAA for use in determining and evaluating an appropriate end play check interval.

2.4.3 Adequacy of End Play Check Procedure

As discussed in the Safety Board's October 1, 2001, safety recommendation letter and its end play data study report, dated March 18, 2002, the current end play check procedure[259] is subject to considerable measurement error. Safety Board investigators observed end play checks performed by maintenance personnel from several MD-80 operators and identified numerous potential sources for inaccurate end play measurements. Specifically, it was noted that the accuracy of the results could be affected by deviations in one or more of the following areas: (1) calibration and interpretation of the dial indicator; (2) installation of the dial indicator; (3) application and direction of the specified torque to the restraining fixture; (4) calibration of the torque wrench; (5) fabrication, lubrication, and maintenance of the restraining fixture; (6) rotation of the acme screw within its gearbox during the procedure; and (7) the individual mechanic's knowledge of the procedures. According to the Board's end play data study report, end play measurement inaccuracies can be as high as 0.030 inch, and errors can occur in either direction.[260]

Further, an additional source of end play measurement error was possible at Alaska Airlines before August 2000 because its fabricated end play restraining fixture was not based on Boeing design specifications.[261] Although there was only one such fixture at Alaska Airlines' maintenance facilities at the time of the accident,[262] 11 additional restraining fixtures were made in-house after the accident despite an April 2000 Boeing message to all DC-9, MD-80, MD-90, and 717 operators asking them to ensure that their horizontal stabilizer inspection tooling conformed "to the tool's drawing requirements." The Alaska Airlines manager of tool control, who participated in the fabrication of the restraining fixtures ordered by airline management after the accident, told Safety Board investigators that "what they were making wasn't even close" to the Boeing drawing. In August 2000, following inquiries from Safety Board staff about which restraining fixture was used during the September 1997 end play check, Alaska Airlines notified the FAA that it was concerned that the restraining fixtures it had manufactured in-house might not be "an equivalent substitute" for the Boeing fixture. Alaska Airlines suspended use of the in-house manufactured restraining fixtures, quarantined all non-Boeing manufactured fixtures, and temporarily removed from service 18 airplanes that had been checked with these fixtures.

The results of Safety Board testing indicated that Alaska Airlines' fabricated restraining fixtures could yield end play measurements that were lower than measurements obtained with a Boeing-manufactured fixture.[263] Therefore, it is possible that the 0.040- and 0.033-inch end play measurements obtained during the accident airplane's September 1997 end play check were less than the actual end play and that the accident jackscrew assembly in fact exceeded the 0.040-inch end play measurement limit

---

[259] As discussed in section 1.6.3.3.2, the end play procedure calls for pulling down on the horizontal stabilizer by applying a specified amount of torque to a tool known as a restraining fixture to change the load on the acme screw from tension to compression. The load reversal allows movement, or play, between the acme screw and nut threads that is measured using a dial indicator, which is mounted on the lower rotational stop (fixed to the screw) with the movable plunger set against the lower surface of the acme nut that measures the relative vertical movement between the acme screw and the nut when the loads are applied. The amount of movement, or end play, is used as an indication of the amount of acme nut thread wear (and acme screw thread wear, if there is any). The procedure calls for the restraining fixture load to be applied and removed several times until consistent measurements are achieved.

[260] For more information, see table 5 in the Safety Board's end play data study report in the public docket for this accident.

[261] For more information about Alaska Airlines' fabricated restraining fixtures, see section 1.6.3.3.3.2.

[262] The Safety Board recognizes that Alaska Airlines contends in its submission that the restraining fixture used during the September 1997 end play check was a Douglas fixture. However, evidence gathered in the Board's investigation does not support this contention.

at the time of the 1997 check. It is also possible that a Boeing-manufactured restraining fixture would have yielded a more accurate reading, which could have resulted in the jackscrew assembly being removed from service. However, as discussed previously, even when a restraining fixture manufactured to Boeing specifications is used, there is still a potential for inaccurate measurements. Therefore, the Safety Board concludes that, until August 2000, Alaska Airlines used a fabricated restraining fixture that did not meet Boeing specifications; however, the Board could not determine whether the use of this noncompliant fixture generated an inaccurate end play measurement during the last end play check or whether the use of this fixture contributed to the accident.

In Safety Recommendations A-01-42 and -44, the Safety Board asked the FAA to do the following:

> Require the Boeing Commercial Airplane Group to revise the end play check procedure for the horizontal stabilizer trim system of Douglas DC-9, McDonnell Douglas MD-80/90, and Boeing 717 series airplanes to minimize the probability of measurement error and conduct a study to empirically validate the revised procedure against an appropriate physical standard of actual acme screw and acme nut wear. This study should also establish that the procedure produces a measurement that is reliable when conducted on-wing. (A-01-42)

> Require maintenance personnel who inspect the horizontal stabilizer trim system of Douglas DC-9, McDonnell Douglas MD-80/90, and Boeing 717 series airplanes to undergo specialized training for this task. This training should include familiarization with the selection, inspection, and proper use of the tooling to perform the end play check. (A-01-44)

In its December 12, 2001, response letter, the FAA indicated that it had asked Boeing to revise the end play check procedure, and the Safety Board is aware that Boeing is currently developing improved end play check procedures. Regarding Safety Recommendation A-01-44, the FAA stated that it believed that "current regulatory requirements of 14 CFR [*Code of Federal Regulations*] 121.375 adequately address maintenance training programs" but that the agency was "revising Advisory Circular [AC] 120-16C, *Continuous Airworthiness Maintenance Programs*, to expand the intent of the requirement for maintenance training programs." In addition, the FAA stated that it believed that "current regulatory requirements of 14 CFR 65.81 and 14 CFR 65.103 adequately address the inspection requirements for maintenance personnel and repairmen." The FAA concluded that "current regulatory requirements in place to require maintenance training programs and inspection requirements for maintenance personnel and repairmen address the full intent of these safety recommendations."

---

[263] Alaska Airlines' fabricated restraining fixture did not result in a lower measurement than the Boeing fixture on every occasion during the testing. However, Safety Board investigators determined that the discrepancies, when they existed, could be as high as 0.005 inch and were likely caused by the fabricated restraining fixture's increased length of thread engagement, which created friction that required more torque than the Boeing-manufactured fixture.

On June 14, 2002, the Safety Board classified Safety Recommendation A-01-42 "Open—Acceptable Response," pending revisions to the end play check procedure. In its letter, the Board classified Safety Recommendation A-01-44 "Open—Unacceptable Response," stating that it "continues to believe that current FAA regulations do not result in maintenance personnel being properly trained to perform lubrication and end play inspection of the acme screw and nut assembly" and urged the FAA to reconsider its decision.

The Safety Board is aware of technologies that could potentially provide more reliable and accurate indications of acme nut thread wear than the end play check procedure. For example, as a result of this accident, Sandia National Laboratories, with FAA funding, began exploring two new methods for monitoring thread wear that involved the use of nondestructive technology. One of these methods used a portable x-ray fixture that could be installed next to the acme screw and nut to provide a graphic depiction of all of the threads. The other method used a small ultrasonic sensor that could be passed over the acme nut to obtain a graphic depiction of the thread crests on a handheld screen.

The Safety Board concludes that the on-wing end play check procedure, as currently practiced, has not been validated and has low reliability. In this regard, the Board will continue to monitor the FAA's and Boeing's actions in response to Safety Recommendations A-01-42 and -44.

2.5 Deficiencies of Jackscrew Assembly Overhaul Procedures and Practices

The accident jackscrew assembly was never overhauled nor was it required to be. However, to determine the adequacy of maintenance and inspection procedures applicable to all DC-9, MD-80/90, and 717 jackscrew assemblies, Safety Board investigators evaluated jackscrew assembly overhaul procedures and practices. Specifically, investigators reviewed the DC-9 Overhaul Maintenance Manual and visited several maintenance facilities that overhaul jackscrew assemblies, including Integrated Aerospace, the only contract facility currently used by Boeing to overhaul acme nut and screw pairs to manufacturing end play specifications.[264]

Safety Board investigators identified several deficiencies in the DC-9 Overhaul Maintenance Manual procedures and in the practices of several of the maintenance facilities that were visited. Specifically, the overhaul manual did not require the use of work cards documenting each step in the overhaul process, and (with the exception of Integrated Aerospace) the facilities did not use such work cards. Also, the manual called for replacement of the acme nut only if the end play measurement was more than 0.040 inch. Although some overhaul facilities had lower self-imposed end play measurement limits for replacement of the acme nut, at least one facility indicated that it

---

[264] For more information about the DC-9 Overhaul Maintenance Manual and Safety Board investigator visits to maintenance facilities, see section 1.18.8.

would return an overhauled jackscrew assembly to a customer as long as the end play measurement did not exceed 0.040 inch. Further, the manual contained no requirement to record or inform the customer of the end play measurement of an overhauled jackscrew assembly. This means that a jackscrew assembly with an end play measurement of up to 0.039 inch could be represented as "overhauled" and returned to service by an operator that might not be aware of the high end play measurement.[265] A jackscrew assembly could require overhaul for reasons other than excessive end play; however, it would be reasonable for a customer to expect that an assembly would be returned after an overhaul with an end play close to manufacturing specifications.

Further, the required steps for properly conducting the end play check procedure were not well described and the required equipment was not specified in the DC-9 Overhaul Maintenance Manual. During visits to maintenance facilities, Safety Board investigators learned that the facilities used different methods and various tools for measuring end play. (Although investigators saw no evidence that the differences affected the accuracy of the results, the lack of standardization nonetheless increases the potential for error to occur.) The overhaul manual also did not contain detailed instructions on how to apply grease to the jackscrew assembly at the completion of the overhaul nor did it clearly specify which type of grease to use. Further, although the manual does require that an overhauled jackscrew assembly be lubricated before it can be returned to the customer, investigators learned that at least one maintenance facility returns overhauled assemblies without lubricating them first.

In addition, the DC-9 Overhaul Maintenance Manual did not contain detailed instructions nor specify appropriate equipment for checking that the proper acme screw thread surface finish had been applied. Many of the maintenance facilities visited by Safety Board investigators indicated that they relied on subvendors to ensure that the proper acme screw thread surface finish was applied and had no standard method for verifying that this action had occurred.

Finally, the DC-9 Overhaul Maintenance Manual did not clearly specify appropriate packaging procedures for transporting jackscrew assemblies after overhaul. Although the overhaul manual contained detailed protective packaging instructions for jackscrew assemblies going into "storage," it did not specify any such protective packaging for jackscrew assemblies being transported. However, it would be prudent to use protective packaging for all overhauled jackscrew assemblies being returned to a customer because a maintenance facility cannot be expected to know what the customer intends to do with an assembly after it is returned.

Integrated Aerospace, the only maintenance facility authorized by Boeing to overhaul the acme nut and screw to an "as new" end play condition that meets

---

[265] Although the DC-9 Overhaul Maintenance Manual calls for more frequent end play checks (every 1,000 flight hours) to be conducted on an overhauled unit that was reinstalled with an end play measurement between 0.034 to 0.039 inch, this provision is of limited value because there is no requirement for operators to be informed of this elevated end play measurement. Further, operators receiving an overhauled jackscrew assembly would not be expected to consult the overhaul manual for special maintenance instructions.

manufacturing specifications, uses more rigorous and reliable overhaul procedures and significant quality control measures that are not used or required by the other facilities that overhaul jackscrew assemblies. For example, Integrated Aerospace uses detailed work cards to document each step of the overhaul process. Whenever Integrated Aerospace receives a jackscrew assembly with an end play measurement greater than 0.015 inch, it will install a new acme nut, thereby restoring the assembly to the manufacturing specifications for a new jackscrew assembly (0.003 to 0.010 inch). In doing so, Integrated Aerospace must comply with detailed specifications provided by Boeing in a service rework drawing, which are not contained in the DC-9 Overhaul Maintenance Manual.

In addition, the investigation revealed that no special authorization beyond a class 1 accessory rating is required for a maintenance facility to overhaul jackscrew assemblies in accordance with the DC-9 Overhaul Maintenance Manual. A class 1 accessory rating allows a facility to perform maintenance and alteration of a number of mechanical accessories. The maintenance facility is not required to demonstrate that it has the necessary capability and equipment to perform jackscrew assembly overhauls. Safety Board investigators found that the PMI of an FAA-certified maintenance facility may not even be aware that the facility is performing overhauls of jackscrew assemblies.

The Safety Board concludes that deficiencies in the overhaul process increase the likelihood that jackscrew assemblies may be improperly overhauled. The Safety Board further concludes that the absence of a requirement to record or inform customers of the end play measurement of an overhauled jackscrew assembly could result in an operator unknowingly returning a jackscrew assembly to service with a higher-than-expected end play measurement. Therefore, the Safety Board believes that the FAA should require that maintenance facilities that overhaul jackscrew assemblies record and inform customers of an overhauled jackscrew assembly's end play measurement.

In addition to recording the end play measurement information provided by a maintenance facility when it returns an overhauled jackscrew assembly, it would also be prudent for operators to record end play measurements for the same assembly after it is installed on an airplane. The Safety Board notes that end play measurements recorded by maintenance facilities are likely to be obtained during bench checks. However, after the overhauled assembly is re-installed on an airplane, end play measurements will likely be obtained through use of the on-wing end play check procedure, which may yield a slightly different measurement because of differences in the procedure. A wear rate that is calculated using a bench-check measurement at one point in time compared with an on-wing measurement at a later point in time will not be as informative or useful as a wear rate that is calculated using measurements obtained through use of the same end play check method. Therefore, the Safety Board concludes that operators will maximize the usefulness of end play measurements and wear rate calculations by recording on-wing end play measurements whenever a jackscrew assembly is replaced on an airplane. Accordingly, the Safety Board believes that the FAA should require operators to measure and record the on-wing end play measurement whenever a jackscrew assembly is replaced.

Finally, the Safety Board concludes that, because the jackscrew assembly is an integral and essential part of the horizontal stabilizer trim system, a critical flight system, it is important to ensure that maintenance facilities authorized to overhaul these assemblies possess the proper qualifications, equipment, and documentation. Therefore, the Safety Board believes that the FAA should require that maintenance facilities that overhaul DC-9, MD-80/90, and 717 series airplanes' jackscrew assemblies obtain specific authorization to perform such overhauls, predicated on demonstrating that they possess the necessary capability, documentation, and equipment for the task and that they have procedures in place to (1) perform and document the detailed steps that must be followed to properly accomplish the end play check procedure and lubrication of the jackscrew assembly, including specification of appropriate tools and grease types; (2) perform and document the appropriate steps for verifying that the proper acme screw thread surface finish has been applied; and (3) ensure that appropriate packaging procedures are followed for all overhauled jackscrew assemblies, regardless of whether the assembly has been designated for storage or shipping.

2.6 Horizontal Stabilizer Trim System Design and Certification Issues

2.6.1 Acme Nut Thread Loss as a Catastrophic Single-Point Failure Mode

The DC-9 horizontal stabilizer trim system (which was also incorporated in the MD-80/90 and 717 series airplanes) is a critical flight system because certain failures of the system can be catastrophic. One such failure is the loss of acme screw and nut thread engagement. However, as previously mentioned, the designers of the system assumed that at least one set of the jackscrew assembly's acme screw and nut threads would always be intact and engaged to act as a load path. Therefore, the repercussions of stripped acme nut threads and the corresponding effect on the airplane (including the possibility of the acme screw disengaging from the acme nut) were not considered in the design of the horizontal stabilizer trim system.

At the Safety Board's public hearing, Boeing engineers stated that they considered loss of the acme nut threads to be a "multiple failure event" and that such loss caused by excessive wear was not considered a "reasonably probable single failure" for certification purposes.[266] The Boeing engineers indicated that the jackscrew assembly was designed to accommodate thread wear but acknowledged that monitoring and managing thread wear was essential to maintaining the integrity of the design. Similarly, an FAA certification

---

[266] The certification basis for the DC-9, MD-80/90, and 717 horizontal stabilizer trim systems was *Civil Aeronautics Regulations* (CAR) 4b. CAR 4b.320, "Control Systems," stated that "an adjustable stabilizer shall incorporate means to permit, after the occurrence of any reasonably probable single failure of the actuating system, such adjustment as would be necessary for continued safety of the flight." (Current certification regulations specify, in 14 CFR 25.671, that the airplane must be shown to be capable of continued safe flight and landing after "any single failure" of the actuating system.)

engineer testified that thread wear "was not considered as a mode of failure for either a systems safety analysis or for structural considerations" and that the "design of the acme nut and screw provided enough over-strength so that regulatory requirements could be met with a significant amount of wear." However, as the Alaska Airlines flight 261 accident demonstrates, complete loss of the acme nut threads because of excessive wear is possible. Further, the accident jackscrew assembly is not the only jackscrew assembly in which excessive acme nut thread wear has occurred. As discussed previously, excessive wear of acme nut threads has occurred on other occasions as a result of inadequate lubrication, improper acme screw thread surface finish, and contamination. In addition, other potential rapid wear mechanisms may not have yet been identified.

The Safety Board notes that the dual-thread design of the acme screw and nut does not adequately protect against excessive acme nut thread wear. Although Boeing contends that the two thread spirals along the length of both the acme screw and nut provide structural redundancy, the Board notes that each set of thread spirals is always carrying loads in flight and that both sets of thread spirals are subject to the same wear mechanisms. Thus, although the dual-thread design may prevent a crack in one thread set from propagating through the other thread set, both sets of threads remain vulnerable to simultaneous wear failure. Therefore, the Safety Board concludes that the dual-thread design of the acme screw and nut does not provide redundancy with regard to wear.

The FAA's certification scheme is intended to protect against catastrophic single-point failure conditions. Specifically, 14 CFR 25.1309 requires that airplane systems and associated components be designed so that "the occurrence of any failure condition which would prevent the continued safe flight and landing of the airplane is extremely improbable." Further, AC 25.1309-1A, "System Design and Analysis," defines "extremely improbable" failure conditions as "those so unlikely that they are not anticipated to occur during the entire operational life of all airplanes of one type" and "having a probability on the order of 1×10^{-9} or less each flight hour, based on a flight of mean duration for the airplane type." AC 25.1309-1A specifies that in demonstrating compliance with 14 CFR 25.1309, "the failure of any single element, component, or connection during any one flight…should be assumed, regardless of its probability," and "such single failures should not prevent continued safe flight and landing, or significantly reduce the capability of the airplane or the ability of the crew to cope with the resulting failure condition."

An FAA senior certification engineer testified at the public hearing that 14 CFR 25.1309[267] did not apply to the jackscrew assembly acme nut because the FAA did not consider it part of a system. Rather, he stated that the jackscrew assembly was a "combination structural element and systems element" and that each category was governed by its respective regulatory requirements. He indicated that those portions of the jackscrew assembly below the gearbox and trim motors carried primary flight loads and, therefore, were covered by regulations pertaining to structure, not systems. He testified that the acme nut met the applicable regulatory requirements pertaining to structure,

---

[267] The predecessor to the current 14 CFR 25.1309 was CAR 4b.606, which required that all systems "be designed to safeguard against hazards to the airplane in the event of their malfunctioning or failure."

specifically, CAR 4b.201(a), "Strength and Deformation," which states that structure "shall be capable of supporting limit loads without suffering detrimental permanent deformation."[268] He stated that acme nut threads that are within manufacturing specifications far exceed the requirements for ultimate strength and deflection limit load and added that the FAA does not consider deflection of worn acme nut threads to be deformation in the context of this regulation.

In addition, a Boeing structures engineering manager testified that the acme nut complied with certification regulations pertaining to fatigue evaluation of structure. Specifically, CAR 4b.270, "Fatigue Evaluation of Flight Structure," required, for "those portions of the airplane's flight structure in which fatigue may be critical," an evaluation of either fatigue strength (also referred to as "safe life") or fail safe strength (also referred to as "damage tolerance").[269] However, the Boeing manager testified that neither evaluation was performed because the acme nut was not considered "fatigue critical" because of its robust design.

It is unclear whether the design and certification of the DC-9 (and MD-80/90 and 717) horizontal stabilizer trim system would have been any different if the certification requirements for aircraft systems, in addition to those applicable to structure, had been applied to the jackscrew assembly acme nut during the design phase. Boeing engineers testified at the public hearing that the horizontal stabilizer trim system design complied with the requirements of 14 CFR 25.1309. However, the FAA certification engineer indicated that, even if section 25.1309 had been applicable to the entire jackscrew assembly, acme nut thread wear would not have been considered in the required systems safety analysis. He explained, "if you refer to a [section 25.1309] type safety analysis to try and determine a failure rate for a wear item, there are not, and there's really no such thing as a wear-critical item, never has been. The question of wear being a quantifiable element so that one could do a safety-type analysis for structure—it's not feasible. The data to do such an evaluation is not available. It doesn't exist."[270]

In sum, the Safety Board is concerned that Boeing and the FAA did not account for the catastrophic effects of total acme nut thread loss in the design and certification of the horizontal stabilizer trim control system. The Board is also concerned that the certification requirements for aircraft systems were not considered applicable to the entire jackscrew assembly, particularly the acme nut. Because the loss of acme nut threads in flight most likely would result in the catastrophic loss of the airplane, the Board considers the acme nut to be a critical element of the horizontal stabilizer trim control system; therefore, it should have been covered by the certification philosophy and regulations applicable to all other flight control systems. The Safety Board concludes that the design of the DC-9, MD-80/90, and 717 horizontal stabilizer jackscrew assembly did not account for the loss of the acme nut threads as a catastrophic single-point failure mode. The Safety Board

---

[268] A substantially similar requirement is currently contained in 14 CFR 25.305(a).

[269] Substantially similar requirements are currently contained in 14 CFR 25.571.

[270] The Boeing manager confirmed that "the condition of wear or wear-out was not included in the original DC-9 fault analysis."

further concludes that the absence of a fail-safe mechanism to prevent the catastrophic effects of total acme nut thread loss contributed to the Alaska Airlines flight 261 accident.

2.6.2 Prevention of Acme Nut Thread Loss Through Maintenance and Inspection

Currently, prevention of acme nut thread loss is dependent on regular application of lubrication and on recurrent inspections of the jackscrew assembly to monitor acme nut thread wear. However, this maintenance-based approach to maintaining the horizontal stabilizer trim system's structural integrity has weaknesses.

First, current lubrication and end play check intervals may not be adequate, and their length can change. Research and testing relating to lubrication effectiveness and the prior service history of the MD-80 fleet suggests that the current 650-flight-hour lubrication interval is probably adequate to ensure proper lubrication of the jackscrew assembly. However, because of the potential for additional undiscovered rapid wear mechanisms and uncertainty regarding the maximum possible wear rate even for known wear mechanisms, there is no such basis for assuming that the current 2,000-flight-hour end play check interval is sufficient to ensure fleetwide safety.[271] Further, there is no guarantee that these intervals will not eventually be extended, which would further reduce the level of safety.

Second, and more importantly, all maintenance and inspection tasks are subject to human error. As previously discussed, this investigation has identified several weaknesses in the lubrication and inspection procedures that could affect their intended results and compromise safety.[272] Further, several Safety Board accident investigations, including the Alaska Airlines flight 261 investigation, have demonstrated that even simple maintenance tasks are sometimes missed or inadequately performed and can have catastrophic results.[273] Therefore, the current horizontal stabilizer trim system design remains

---

[271] For a more detailed discussion of the Safety Board's concerns in this area, see section 2.4.2.

[272] For more information about the weaknesses in the lubrication and inspection procedures, see sections 2.3.6.3 and 2.4.3, respectively.

[273] For example, following the September 11, 1991, crash of a Continental Express Embraer 120 in Eagle Lake, Texas, the Safety Board concluded that the airline's maintenance inspection and quality assurance programs failed to detect that the upper row of screws on the leading edge of the left horizontal stabilizer had been removed during maintenance and had not been replaced. The partially secured left horizontal stabilizer leading edge separated in flight, causing a severe nose-down pitchover. The airplane broke up in flight, and all 14 people on board were killed. For more information, see *National Transportation Safety Board. Britt Airways, Inc., d/b/a Continental Express Flight 2574, In-flight Structural Breakup, EMB-120RT, N33701, Eagle Lake, Texas, September 11, 1991.* NTSB/AAR-92/04. In addition, the Board's investigation of the July 6, 1996, uncontained engine failure on a Delta Air Lines MD-88 in Pensacola, Florida, determined that a fluorescent inspection process used to detect fatigue cracks during maintenance was susceptible to error because it involved multiple cleaning, processing, and inspection procedures dependent on several individuals and because of a low expectation of finding a crack. Shrapnel from the uncontained engine failure pierced the fuselage and entered the rear cabin. Two passengers were killed, and two others were seriously injured. For more information, see *National Transportation Safety Board. Uncontained Engine Failure, Delta Airlines Flight 1288, McDonnell Douglas MD-88, N927DA, Pensacola, Florida, July 6, 1996.* NTSB/AAR-98-01.

vulnerable to catastrophic failure if maintenance and inspection tasks are not performed properly.

2.6.3 Elimination of Catastrophic Effects of Acme Nut Thread Loss Through Design

The Safety Board concludes that, when a single failure could have catastrophic results and there is a practicable design alternative that could eliminate the catastrophic effects of the failure mode, it is not appropriate to rely solely on maintenance and inspection intervention to prevent the failure from occurring; if a practicable design alternative does not exist, a comprehensive systemic maintenance and inspection process is necessary. In the case of the horizontal stabilizer trim system, such a design would incorporate a reliable, independent means for eliminating, overcoming, or counteracting the catastrophic effects of acme nut thread loss. The Safety Board notes that such a design change would not necessarily need to incorporate dual actuators, or any other form of system redundancy; the design would only need to provide a mechanism for preventing stripped acme nut threads from resulting in unrecoverable movement of the horizontal stabilizer. The Board notes that among the several design concepts listed in AC 25.1309-1A that can be used to avoid catastrophic failure conditions are the following: (1) "designed failure effect limits, including the capability to sustain damage, to limit the safety impact or effects of a failure"; and (2) "designed failure path to control and direct the effects of a failure in a way that limits its safety impact."

The Safety Board concludes that transport-category airplanes should be modified, if practicable, to ensure that horizontal stabilizer trim system failures do not preclude continued safe flight and landing. Therefore, the Safety Board believes that the FAA should conduct a systematic engineering review to (1) identify means to eliminate the catastrophic effects of total acme nut thread failure in the horizontal stabilizer trim system jackscrew assembly in DC-9, MD-80/90, and 717 series airplanes and require, if practicable, that such fail-safe mechanisms be incorporated in the design of all existing and future DC-9, MD-80/90, and 717 series airplanes and their derivatives; (2) evaluate the horizontal stabilizer trim systems of all other transport-category airplanes to identify any designs that have a catastrophic single-point failure mode and, for any such system; (3) identify means to eliminate the catastrophic effects of that single-point failure mode and, if practicable, require that such fail-safe mechanisms be incorporated in the design of all existing and future airplanes that are equipped with such horizontal stabilizer trim systems.

Further, the Safety Board is concerned that the FAA certified a horizontal stabilizer trim system that had a single-point catastrophic failure mode. The Safety Board concludes that catastrophic single-point failure modes should be prohibited in the design of all future airplanes with horizontal stabilizer trim systems, regardless of whether any element of that system is considered structure rather than system or is otherwise considered exempt from certification standards for systems. Therefore, the Safety Board believes that the FAA should modify the certification regulations, policies, or procedures to ensure that new horizontal stabilizer trim control system designs are not certified if they have a single-point catastrophic failure mode, regardless of whether any element of that system is considered structure rather than system or is otherwise considered exempt from certification standards for systems.

2.6.4 Consideration of Wear-Related Failures During Design and Certification

The Alaska Airlines flight 261 investigation revealed that the FAA certification processes and procedures did not adequately consider and address the consequences of excessive wear in the context of certifying the DC-9, MD-80/90, and 717 horizontal stabilizer trim system. In light of this finding, the Safety Board is concerned that the consequences of excessive wear might not be considered in the contexts of other certifications as well. One way to ensure that such consequences are considered would be to include wear-related failures in the failure modes and effects analyses (FMEA) and fault tree analyses that are required under 14 CFR 25.1309. As stated previously, Boeing and the FAA have accepted the premise that wear cannot be considered a mode of failure in systems safety analyses such as FMEAs and fault trees; however, the Safety Board notes that standards developed by the Society of Automotive Engineers (SAE) specify that wear should be considered in FMEAs.[274] Design guidelines should require that a wear-related failure be assumed and that the results of such a failure be evaluated.

The Safety Board concludes that the certification requirements applicable to transport-category airplanes should fully consider and address the consequences of failures resulting from wear. Therefore, the Safety Board believes that the FAA should review and revise aircraft certification regulations and associated guidance applicable to the certification of transport-category airplanes to ensure that wear-related failures are fully considered and addressed so that, to the maximum extent possible, they will not be catastrophic.

2.7 Deficiencies in Alaska Airlines' Maintenance Program

2.7.1 April 2000 FAA Special Inspection Findings

As a result of the Alaska Airlines flight 261 accident, the FAA conducted a special inspection of Alaska Airlines in April 2000, which was headed by a member of its System Process Audit staff from Washington, D.C. The FAA's final report on this special inspection, dated June 20, 2000, stated the following:

- procedures in place at the company are not being followed;

---

[274] SAE's Aerospace Recommended Practices 4761, "Guidelines and Methods for Conducting the Safety Assessment Process on Civil Airborne Systems and Equipment," in paragraph G.3.2.2.1 of appendix G, lists wear among the failure modes to consider in performing an FMEA.

- controls in place are clearly not effective;
- authority and responsibilities are not well defined;
- control of the deferral systems is missing; and
- quality control and quality assurance programs are ineffective.

The report also noted that the continuing analysis and surveillance program was responsible for "overseeing the operations within the company to ensure that the programs established…are effective" but that because Alaska Airlines did not have a "functional [continuing analysis and surveillance program], numerous other areas suffer from the lack of oversight and reform." The report concluded that most of the findings could be attributed to Alaska Airlines' processes, or the lack thereof, and "ineffective quality control and quality assurance departments."[275]

Specifically, the special inspection report revealed that the Alaska Airlines General Maintenance Manual (GMM) did "not reflect the procedures that the company is actually using to perform maintenance on its aircraft at the company's maintenance facilities." The report documented inadequacies in the GMM procedures for issuance of airworthiness releases of airplanes coming out of heavy checks, incomplete C-check paperwork, discrepancies of shelf-life expiration dates of consumables, a lack of engineering approval and quality control of work card modifications in heavy checks, and inadequate tool calibrations.

The special inspection report also stated that the GMM did not "specify maintenance training curriculums or on-the-job training (OJT) procedures or objectives" and cited 14 CFR 121.375, which requires that Part 121 operators have a training program to ensure competency. The report noted that the OJT program was "informal [and] administered at the discretion of the appointed instructors. The program is not structured, there is no identification of subjects to be covered, and there are no criteria for successful completion provided."

As a result of the special inspection team's findings, the FAA initially proposed the suspension of Alaska Airlines' authority to conduct heavy maintenance. Instead, the FAA accepted an Airworthiness and Operations Action Plan submitted by Alaska Airlines and did not suspend Alaska Airlines' heavy maintenance authority. In September 2000, the FAA conducted a followup inspection, which revealed that Alaska Airlines had not satisfactorily resolved many of the deficiencies identified in the special inspection report. After an additional review in July 2001, consisting of presentations by Alaska Airlines representatives and a brief visit by some of the panel members to Alaska Airlines' SEA maintenance facility followed by discussions about Alaska Airlines' progress, the FAA found that the airline had "met or exceeded all commitments set forth in their action plan" and "demonstrated…that previously identified systemic deficiencies have been corrected." The FAA further found that Alaska Airlines had made "significant changes and improvements."

---

[275] For more detailed information about the FAA special inspection team's findings, see section 1.17.3.4.

2.7.2 Maintenance-Related Deficiencies Identified During This Accident Investigation

In the course of its investigation of the Alaska Airlines flight 261 accident, the Safety Board learned about numerous maintenance-related actions—both general and specific—by Alaska Airlines management and maintenance personnel. These actions supported the overall findings in the FAA's special inspection report and were indicative of procedural and quality control deficiencies, which concerned the Board.

2.7.2.1 General Policy Decisions

Extension of End Play Check Interval

As previously discussed, in 1996, Alaska Airlines extended its C-check interval from 13 to 15 months, which resulted in the end play check interval being increased from 26 to 30 months. Although 30 months was the manufacturer-recommended calendar-time interval, because of Alaska Airlines' increased airplane utilization rate, 30 months was equivalent to 9,550 flight hours, which was well beyond the 7,000- or 7,200-flight-hour interval that was recommended by Boeing at the time.[276] As previously noted, the flight 261 accident occurred almost 9,000 flight hours after the accident jackscrew assembly's last end play check. Therefore, if Alaska Airlines had adopted an interval that complied with either of the flight-hour intervals specified by Boeing, the accident jackscrew assembly would have undergone an end play check 1,800 to 2,000 flight hours before the accident occurred and might have been removed at that time for having an excessive end play measurement. The decision to exceed the manufacturer's recommended end play check interval was a poor maintenance decision and demonstrated a lack of appreciation for the need to account for flight hours when setting maintenance intervals for those tasks related to airplane usage rather than calendar time.

Grease Change from Mobilgrease 28 to Aeroshell 33

The process by which Alaska Airlines replaced Mobilgrease 28 with Aeroshell 33 grease on its work maintenance task cards for the MD-80 was both procedurally and substantively deficient in several respects.

First, the grease change was accomplished without following established procedures. The form used to request and obtain internal approval for the grease change was incomplete and inadequate in the following ways: (1) it was missing several required signatures;[277] (2) it contained no indication that the required internal approval from Alaska Airlines' Reliability Analysis Program Control Board had been obtained; and (3) it contained no signature on the line labeled "change request accomplished," although the

---

[276] As previously mentioned, MSG-2 guidance (which was the basis for Alaska Airlines' maintenance program) specified an end play check interval of 7,000 flight hours, and the more recent MSG-3 guidance specified an interval of 7,200 flight hours.

[277] There was no signature from the manager of reliability, the director of line maintenance, or the manager of quality control.

grease change was implemented in December 1997. At the Safety Board's public hearing, neither Alaska Airlines' former director of engineering (who at the time the change was processed was Alaska Airlines' manager of maintenance programs and technical publications) nor the current director of engineering could explain how the change was implemented without the required signatures or Reliability Analysis Program Control Board approval. Further, the GMM was not modified to reflect the grease change to Aeroshell 33; it still specified the use of Mobilgrease 28. This lapse indicates either a lack of procedures or a failure to follow procedures.

Second, Alaska Airlines did not conduct any formal followup to evaluate the performance of the new grease (Aeroshell 33). McDonnell Douglas responded to Alaska Airlines' request to change greases in a letter stating that it had "no technical objection" to the use of Aeroshell 33; however, it also stated that McDonnell Douglas could not "yet verify the performance of this grease" and that it would "be the responsibility of Alaska Airlines to monitor the areas where Aeroshell 33 grease is used for any adverse reactions." However, Alaska Airlines did not actively monitor the in-service performance of the grease. Although Alaska Airlines asserts that it monitored the performance of Aeroshell 33 through its overall reliability program, the Safety Board notes that there is no evidence that any specific data were collected and evaluated regarding the lubricating effectiveness of Aeroshell 33. For example, it would have been prudent for Alaska Airlines to measure and record end play and calculate the wear rate on a sample of airplanes after the introduction of Aeroshell 33 and perhaps even to shorten the lubrication intervals in order to verify that its performance was at least equivalent to that of Mobilgrease 28. Alaska Airlines' failure to actively monitor the performance of the new grease is especially significant in light of the extension in lubrication intervals for the horizontal stabilizer components (from 1,600 to 2,550 flight hours) that it had implemented the year before.

Third, Alaska Airlines' technical justification for the grease change was inadequate. When Alaska Airlines notified the FAA in December 1997 of the grease change to Aeroshell 33, it did not provide any substantiating justification at that time nor was such action required. However, when the FAA requested substantiating justification after concerns were raised in connection with the flight 261 investigation, the only documents Alaska Airlines submitted to the FAA were a trade magazine article on Aeroshell 33, excerpts from Boeing 737 and MD-80 maintenance manuals, Boeing service letters, internal correspondence and messages between Alaska Airlines and Boeing, and existing MIL specifications for Aeroshell 33 and Mobilgrease 28. None of these documents contained any information or performance data specifically applicable to the use of Aeroshell 33 on MD-80 airplanes or on the horizontal stabilizer systems of such airplanes. These documents were apparently the only basis Alaska Airlines had for justifying the grease change internally.[278]

---

[278] Alaska Airlines officials did not offer any of the additional information that they claim was used to show internal justification for the grease change. Alaska Airlines' director of engineering testified at the hearing that he thought that the documents submitted to the FAA provided sufficient justification for the change.

The FAA determined that the substantiating documents submitted by Alaska Airlines did not support the change and it disapproved the use of Aeroshell Grease 33 as a substitute for Mobilgrease 28. The Safety Board agrees that the documents submitted to the FAA did not adequately justify the grease change. Even though the use of Aeroshell 33 for lubrication of the horizontal stabilizer system was not determined to be a factor in the Alaska Airlines flight 261 accident, the Board notes that Alaska Airlines disregarded the potential safety impact of a change in grease.

Failure to Fill Required Senior Positions

As noted in the FAA special inspection report, three required positions—the director of maintenance, the director of operations, and the director of safety—were not filled at the time of the special inspection.[279] Although the duties of the director of maintenance were being performed by two people, the division of duties and responsibilities was not clearly defined. The report noted, "consequently there is confusion as to who is responsible for what tasks." The report also noted that the director of safety did not report "directly to the highest level of management." The lack of a full-time director of maintenance and director of safety indicates a lack of sufficient emphasis on maintenance and safety.

Use of Nonconforming Restraining Fixtures

At the time of the accident, Alaska Airlines had only one restraining fixture in its inventory (at its OAK maintenance facility). The fixture was manufactured in-house and it did not meet the manufacturer's required specifications for this fixture. About 1 month after the accident, Alaska Airlines manufactured 11 additional restraining fixtures similar in design to its original in-house manufactured fixture. Alaska Airlines' manager of tool control, who participated in the fabrication of the fixtures ordered by airline management after the accident, told Safety Board investigators in an interview that "what they were making [the restraining fixtures] wasn't even close" to the Boeing drawing. He added that "we were directed to build the tools, and we did exactly what we were told."

On August 2, 2000, in response to Safety Board queries regarding the physical properties of the in-house manufactured restraining fixtures, Alaska Airlines notified the FAA that the restraining fixtures might not be "an equivalent substitute" for the Boeing fixture and, therefore, could produce erroneous measurements. Alaska Airlines thereafter suspended use of its in-house manufactured restraining fixtures and temporarily grounded 18 of its 34 MD-80 series airplanes after determining that these airplanes may have received end play checks with the in-house manufactured fixtures.[280]

Although use of Alaska Airlines' in-house manufactured restraining fixture to perform the accident airplane's last end play check might not have contributed to the

---

[279] Title 14 CFR 119.65 requires airlines to have full-time, qualified personnel in each of these positions.

[280] Alaska Airlines performed end play checks on the grounded airplanes using Boeing-manufactured restraining fixtures and found that they had acceptable end play measurements.

Alaska Airlines flight 261 accident, the use of this nonconforming fixture is further evidence of a flawed maintenance program.

2.7.2.2 Specific Maintenance Actions

In addition to the general policy-related deficiencies outlined above, the Alaska Airlines flight 261 accident investigation also revealed several individual maintenance actions indicative of poor adherence to procedures and/or poor quality control.

Inadequate Lubrication of Jackscrew Assemblies

The investigation revealed that the jackscrew assemblies from three of Alaska Airlines 34 MD-80 airplanes (the accident assembly and those from N981AS and N982AS discovered in February 2000) had excessively high wear rates. The high wear rates on these jackscrew assemblies could not be explained by contamination, improper acme thread surface finish, or anything other than a lack of lubrication. As previously discussed, at least one lubrication had to have been missed or inadequately performed on the accident assembly.[281] However, a scenario in which more than one lubrication was missed or inadequate results in a wear rate more consistent with that which Safety Board testing showed to be the wear rate of unlubricated test specimens. Using the same reasoning, at least one lubrication, and possibly more, must also have been missed entirely or inadequately performed on the jackscrew assemblies on N981AS and N982AS. This indicates either poor training in this task or poor supervision of maintenance personnel performing this task.

Inadequate Lubrication of Rudder Trim Tab Hinge Support Bearings

On July 19, 2002, Alaska Airlines issued a Maintenance Information Letter stating that "[r]ecent shop findings have revealed that the MD-80 rudder trim tab support bearings may not always be receiving adequate lubrication." The letter indicated that two airplanes that had recently come out of a C check, during which the bearings should have been lubricated, showed "extreme bearing wear, with little or no evidence of grease" and that one of the bearings "fell apart upon removal." Alaska Airlines subsequently indicated that, contrary to what was stated in the letter, the airplanes had not recently undergone a C check. However, regardless of how long it had been since the last C check or scheduled lubrication, the condition of these bearings strongly suggests that they were not adequately lubricated at the last scheduled lubrication opportunity. This raises significant concerns about Alaska Airlines' lubrication practices, especially in light of the findings from this investigation relating to inadequate lubrication of jackscrew assemblies.

Failure to Properly Process Order for New Jackscrew Assembly

Following the 0.040-inch end play measurement on the accident jackscrew assembly on September 27, 1997, a nonroutine work card (MIG-4) was generated to begin

---

[281] For more information on the analysis of how many lubrications might have been missed or inadequately performed, see section 2.3.6.1.

the process of ordering a replacement jackscrew assembly. The "planned action" noted on the nonroutine work card was "Replace nut and perform E.O. [engineering order] 8-55-10-01." However, a new jackscrew assembly was never ordered. According to Alaska Airlines' director of safety, Alaska Airlines' procedures called for the creation of a field requisition to order a nonstocked part, such as a jackscrew assembly, but no such requisition was found. Further, the E.O. referenced in the nonroutine work card (which was also referenced in the Alaska Airlines maintenance task card) was incorrect because it applied to another airplane model.

The Safety Board notes that 3 days after the 0.040-inch end play measurement was recorded, a 0.033-inch measurement was recorded, and the accident airplane was thereafter returned to service. The Board recognizes that there was no requirement for Alaska Airlines to order a new jackscrew assembly because it was permissible to return the assembly to service with a recorded end play measurement of either 0.040 or 0.033 inch. Nonetheless, the failure of Alaska Airlines maintenance personnel to process the order for a new part is yet another example of failure to comply with internal procedures. Further, the incorrect E.O. reference is an additional example of poor maintenance quality control.

Continued Errors in Performing End Play Checks

Finally, the Safety Board notes that even after the attention that has been focused on the end play check procedure as a result of the flight 261 accident, Alaska Airlines maintenance personnel continue to have difficulty properly performing the procedure. In March and April 2002, Alaska Airlines removed two jackscrew assemblies, one at OAK and one at a contract facility near SEA, after reported end play measurements of only 0.001 inch. The manufacturing specifications require a minimum end play of 0.003 inch; therefore, a lower end play measurement is suspect. Safety Board investigators examined these jackscrew assemblies and determined that their end play measurements were within normal and expected limits and that the 0.001-inch end play measurements were caused by maintenance errors in performing the end play check procedure. Specifically, it was determined that such inaccurate measurements would result from applying torque to the restraining fixture in the wrong direction before measuring the end play.

According to Boeing, during the 6-month period from December 2001 to May 2002, only six other end play measurements of less than 0.004 inch were reported. Therefore, it appears that Alaska Airlines has experienced a relatively high percentage of such erroneous readings. Further, Alaska Airlines indicated that all maintenance personnel who would be performing end play checks had undergone 4 hours of specialized training on how to properly conduct the end play check procedure. It is noteworthy that, given the heightened awareness resulting from the Alaska Airlines flight 261 accident, and even with this additional training, end play check procedural errors continued to occur at Alaska Airlines.

2.7.3 Summary

On the basis of the information in sections 2.7.2.1 and 2.7.2.2, the Safety Board concludes that at the time of the flight 261 accident, Alaska Airlines' maintenance program had widespread systemic deficiencies. As previously discussed, two specific maintenance-related deficiencies—insufficient lubrication of the accident jackscrew assembly and extension of the end play check interval—were causal to the accident. The Board could not establish whether any of the other maintenance-related deficiencies discovered in connection with this accident investigation were directly related to, or contributed to, the accident. However, these deficiencies are significant departures from expected and established safety standards and, therefore, could well have compromised safety in other ways.

Alaska Airlines has made numerous changes in response to the FAA's postaccident special inspection findings, and an FAA panel has indicated that it is satisfied that the "previously identified systemic deficiencies have been corrected." Nonetheless, the Safety Board is concerned that these widespread deficiencies existed at Alaska Airlines before the accident and were not identified or corrected until after attention was focused on them as a result of this accident investigation. The Board is also concerned, in light of the recent maintenance errors during end play check procedures and evidence of continued poor lubrication practices, that the deficiencies at Alaska Airlines may not have been fully corrected.

2.8 FAA Oversight

The FAA's April 2000 special inspection uncovered widespread significant deficiencies that the FAA should have identified earlier. The fact that these deficiencies were not discovered until special attention was focused on Alaska Airlines by a headquarters-led team as result of the flight 261 accident indicates that the FAA's surveillance was ineffective before the accident.

Several FAA officials acknowledged that surveillance of Alaska Airlines had been inadequate for at least a year before the accident. The former PMI for Alaska Airlines testified that the replacement of the Program Tracking and Reporting System (PTRS) of oversight at Alaska Airlines with the Air Transportation Oversight System (ATOS) in October 1998 resulted in "a terrible transition" that drastically reduced the amount of time inspectors had for actual surveillance activities. He indicated that as a result of additional workload and other administrative changes associated with the implementation of ATOS, "nobody was out there looking at the carrier."[282]

The Seattle Flight Standards District Office (FSDO) CMO supervisor confirmed that "the amount of surveillance that we have done since the introduction of ATOS has probably generally decreased." His concern about this decrease in oversight is reflected in

---

[282] For more information about the impact of the implementation of ATOS on the FAA's oversight of Alaska Airlines, see section 1.17.3.3.

a November 12, 1999, memorandum that he sent to the FAA's Flight Standards Division stating that additional staffing would be needed for the Seattle FSDO CMO to carry out its assigned duties. The memorandum stated the following:

> [staffing at the Seattle FSDO CMO] has reached a critical point...we are not able to properly meet the workload demands. Alaska Airlines has expressed continued concern over our inability to serve it in a timely manner. Some program approvals have been delayed or accomplished in a rushed manner at the 'eleventh hour' and we anticipate this problem will intensify with time. Also, many enforcement investigations, particularly in the area of cabin safety, have been delayed as a result of resource shortages...in order for the Seattle FSDO [CMO] to accommodate the significant volume of Alaska Airline's 'demand' work and effectively meet the objectives of the ATOS program, additional inspector staffing must be made available.

The memorandum also indicated that if the Seattle FSDO CMO "continues to operate with the existing limited number of airworthiness inspectors...diminished surveillance is imminent and the risk of incidents or accidents at Alaska Airlines is heightened."

Therefore, the Safety Board concludes that the FAA did not fulfill its responsibility to properly oversee the maintenance operations at Alaska Airlines, and that at the time of the Alaska Airlines flight 261 accident, FAA surveillance of Alaska Airlines had been deficient for at least several years. The Safety Board notes that after the accident, PTRS-based oversight of Alaska Airlines was re-instituted, and staffing at the CMO was increased significantly. Therefore, it appears that the FAA has taken steps to increase the level of its surveillance of Alaska Airlines. In July 2001, an FAA panel determined that Alaska Airlines had corrected the previously identified deficiencies. However, in light of (1) the systemic nature and widespread impact of those deficiencies, (2) Alaska Airlines' failure to demonstrate during the FAA's first followup inspection (in September 2000) that it had fully corrected the deficiencies, (3) the absence of any in-depth followup inspection by an objective team of FAA inspectors not personally responsible for the oversight of Alaska Airlines to verify that the deficiencies have in fact been corrected,[283] and (4) recent maintenance errors by Alaska Airlines maintenance personnel during end play check procedures and evidence of continued poor lubrication practices,[284] the Board questions the depth and effectiveness of Alaska Airlines' corrective actions and remains concerned about the overall adequacy of Alaska Airlines' maintenance program.

---

[283] The FAA panel's conclusion that the deficiencies had been corrected was based primarily on briefings from Alaska Airlines officials.

[284] The Safety Board notes that these maintenance errors seem to underscore the need for specialized training for personnel who lubricate and perform end play checks, as the Board recommended in Safety Recommendations A-01-43 and -44. Those recommendations are currently classified "Open—Unacceptable Response."

3. Conclusions

3.1 Findings

1. The flight crewmembers on Alaska Airlines flight 261 were properly certificated and qualified and had received the training and off-duty time prescribed by Federal regulations. No evidence indicated any preexisting medical or other condition that might have adversely affected the flight crew's performance during the accident flight.

2. The airplane was dispatched in accordance with Federal Aviation Administration regulations and approved Alaska Airlines procedures. The weight and balance of the airplane were within limits for dispatch, takeoff, climb, and cruise.

3. Weather was not a factor in the accident.

4. There was no evidence of a fire or of impact with birds or any other foreign object.

5. No evidence indicated that the airplane experienced any preimpact structural or system failures, other than those associated with the longitudinal trim control system, the horizontal stabilizer, and its surrounding structure.

6. Both engines were operating normally before the final dive.

7. Air traffic control personnel involved with the accident flight were properly certificated and qualified for their assigned duty stations.

8. The longitudinal trim control system on the accident airplane was functioning normally during the initial phase of the accident flight.

9. The horizontal stabilizer stopped responding to autopilot and pilot commands after the airplane passed through 23,400 feet. The pilots recognized that the longitudinal trim control system was jammed, but neither they nor the Alaska Airlines maintenance personnel could determine the cause of the jam.

10. The worn threads inside the horizontal stabilizer acme nut were incrementally sheared off by the acme screw and were completely sheared off during the accident flight. As the airplane passed through 23,400 feet, the acme screw and nut jammed, preventing further movement of the horizontal stabilizer until the initial dive.

11. The accident airplane's initial dive from 31,050 feet began when the jam between the acme screw and nut was overcome as a result of operation of the primary trim motor. Release of the jam allowed the acme screw to pull up through the acme nut, causing the horizontal stabilizer leading edge to move upward, thus causing the airplane to pitch rapidly downward.

12. The acme screw did not completely separate from the acme nut during the initial dive because the screw's lower mechanical stop was restrained by the lower surface of the acme nut until just before the second and final dive about 10 minutes later.

13. The cause of the final dive was the low-cycle fatigue fracture of the torque tube, followed by the failure of the vertical stabilizer tip fairing brackets, which allowed the horizontal stabilizer leading edge to move upward significantly beyond what is permitted by a normally operating jackscrew assembly. The resulting upward movement of the horizontal stabilizer leading edge created an excessive upward aerodynamic tail load, which caused an uncontrollable downward pitching of the airplane from which recovery was not possible.

14. In light of the absence of a checklist requirement to land as soon as possible and the circumstances confronting the flight crew, the flight crew's decision not to return to Lic Gustavo Diaz Ordaz International Airport, Puerto Vallarta, Mexico, immediately after recognizing the horizontal stabilizer trim system malfunction was understandable.

15. The flight crew's decision to divert the flight to Los Angeles International Airport, Los Angeles, California, rather than continue to San Francisco International Airport, San Francisco, California, as originally planned was prudent and appropriate.

16. Alaska Airlines dispatch personnel appear to have attempted to influence the flight crew to continue to San Francisco International Airport, San Francisco, California, instead of diverting to Los Angeles International Airport, Los Angeles, California.

17. The flight crew's use of the autopilot while the horizontal stabilizer was jammed was not appropriate.

18. Flight crews dealing with an in-flight control problem should maintain any configuration change that would aid in accomplishing a safe approach and landing, unless that configuration change adversely affects the airplane's controllability.

19. Without clearer guidance to flight crews regarding which actions are appropriate and which are inappropriate in the event of an inoperative or malfunctioning flight control system, pilots may experiment with improvised troubleshooting measures that could inadvertently worsen the condition of a controllable airplane.

20. The acme nut threads on the accident airplane's horizontal stabilizer jackscrew assembly wore at an excessive rate.

21. Alaska Airlines' use of Aeroshell 33 for lubrication of the jackscrew assembly, acme screw thread surface finish, foreign debris, and abnormal loading of the acme nut threads were not factors in the excessive wear of the accident acme nut threads.

22. There was no effective lubrication on the acme screw and nut interface at the time of the Alaska Airlines flight 261 accident.

23. The excessive and accelerated wear of the accident jackscrew assembly acme nut threads was the result of insufficient lubrication, which was directly causal to the Alaska Airlines flight 261 accident.

24. Alaska Airlines' extensions of its lubrication interval for its McDonnell Douglas MD-80 horizontal stabilizer components and the Federal Aviation Administration's approval of these extensions, the last of which was based on Boeing's extension of the recommended lubrication interval, increased the likelihood that a missed or inadequate lubrication would result in excessive wear of jackscrew assembly acme nut threads and, therefore, was a direct cause of the excessive wear and contributed to the Alaska Airlines flight 261 accident.

25. When lubricating the jackscrew assembly, removal of used grease from the acme screw before application of fresh grease will increase the effectiveness of the lubrication.

26. A larger access panel would facilitate the proper accomplishment of the jackscrew assembly lubrication task.

27. If the jackscrew assembly lubrication procedure were a required inspection item for which an inspector's signoff is needed, the potential for unperformed or improperly performed lubrications would be reduced.

28. Alaska Airlines' extension of the end play check interval and the Federal Aviation Administration's approval of that extension allowed the accident acme nut threads to wear to failure without the opportunity for detection and, therefore, was a direct cause of the excessive wear and contributed to the Alaska Airlines flight 261 accident.

29. Alaska Airlines' end play check interval extension should have been, but was not, supported by adequate technical data to demonstrate that the extension would not present a potential hazard.

30. The existing process by which manufacturers revise recommended maintenance task intervals and by which airlines establish and revise these intervals does not include task-by-task engineering analysis and justification and, therefore, allows for the possibility of inappropriate interval extensions for potentially critical maintenance tasks.

31. Because of the possibility that higher-than-expected wear could cause excessive wear in less than 2,000 flight hours and the additional possibility that an end play check could be not performed or improperly performed, the current 2,000-flight-hour end play check interval specified in Airworthiness Directive 2000-15-15 may be inadequate to ensure the safety of the Douglas DC-9, McDonnell Douglas MD-80/90, and Boeing 717 fleet.

32. The continued collection and analysis of end play data are critical to monitoring acme nut thread wear and identifying excessive or unexpected wear rates, trends, or anomalies.

33. Until August 2000, Alaska Airlines used a fabricated restraining fixture that did not meet Boeing specifications; however, the Safety Board could not determine whether the use of this noncompliant fixture generated an inaccurate end play measurement during the last end play check or whether the use of this fixture contributed to the accident.

34. The on-wing end play check procedure, as currently practiced, has not been validated and has low reliability.

35. Deficiencies in the overhaul process increase the likelihood that jackscrew assemblies may be improperly overhauled.

36. The absence of a requirement to record or inform customers of the end play measurement of an overhauled jackscrew assembly could result in an operator unknowingly returning a jackscrew assembly to service with a higher-than-expected end play measurement.

37. Operators will maximize the usefulness of end play measurements and wear rate calculations by recording on-wing end play measurements whenever a jackscrew assembly is replaced on an airplane.

38. Because the jackscrew assembly is an integral and essential part of the horizontal stabilizer trim system, a critical flight system, it is important to ensure that maintenance facilities authorized to overhaul these assemblies possess the proper qualifications, equipment, and documentation.

39. The dual-thread design of the acme screw and nut does not provide redundancy with regard to wear.

40. The design of the Douglas DC-9, McDonnell Douglas MD-80/90, and Boeing 717 horizontal stabilizer jackscrew assembly did not account for the loss of the acme nut threads as a catastrophic single-point failure mode. The absence of a fail-safe mechanism to prevent the catastrophic effects of total acme nut thread loss contributed to the Alaska Airlines flight 261 accident.

41. When a single failure could have catastrophic results and there is a practicable design alternative that could eliminate the catastrophic effects of the failure mode, it is not appropriate to rely solely on maintenance and inspection intervention to prevent the failure from occurring; if a practicable design alternative does not exist, a comprehensive systemic maintenance and inspection process is necessary.

42. Transport-category airplanes should be modified, if practicable, to ensure that horizontal stabilizer trim system failures do not preclude continued safe flight and landing.

43. Catastrophic single-point failure modes should be prohibited in the design of all future airplanes with horizontal stabilizer trim systems, regardless of whether any element of that system is considered structure rather than system or is otherwise considered exempt from certification standards for systems.

44. The certification requirements applicable to transport-category airplanes should fully consider and address the consequences of failures resulting from wear.

45. At the time of the flight 261 accident, Alaska Airlines' maintenance program had widespread systemic deficiencies.

46. The Federal Aviation Administration (FAA) did not fulfill its responsibility to properly oversee the maintenance operations at Alaska Airlines, and at the time of the Alaska Airlines flight 261 accident, FAA surveillance of Alaska Airlines had been deficient for at least several years.

3.2 Probable Cause

The National Transportation Safety Board determines that the probable cause of this accident was a loss of airplane pitch control resulting from the in-flight failure of the horizontal stabilizer trim system jackscrew assembly's acme nut threads. The thread failure was caused by excessive wear resulting from Alaska Airlines' insufficient lubrication of the jackscrew assembly.

Contributing to the accident were Alaska Airlines' extended lubrication interval and the Federal Aviation Administration's (FAA) approval of that extension, which increased the likelihood that a missed or inadequate lubrication would result in excessive wear of the acme nut threads, and Alaska Airlines' extended end play check interval and the FAA's approval of that extension, which allowed the excessive wear of the acme nut threads to progress to failure without the opportunity for detection. Also contributing to the accident was the absence on the McDonnell Douglas MD-80 of a fail-safe mechanism to prevent the catastrophic effects of total acme nut thread loss.

4. Recommendations

4.1 New Recommendations

As a result of the investigation of the Alaska Airlines flight 261 accident, the National Transportation Safety Board makes the following recommendations to the Federal Aviation Administration:

Issue a flight standards information bulletin directing air carriers to instruct pilots that in the event of an inoperative or malfunctioning flight control system, if the airplane is controllable they should complete only the applicable checklist procedures and should not attempt any corrective actions beyond those specified. In particular, in the event of an inoperative or malfunctioning horizontal stabilizer trim control system, after a final determination has been made in accordance with the applicable checklist that both the primary and alternate trim systems are inoperative, neither the primary nor the alternate trim motor should be activated, either by engaging the autopilot or using any other trim control switch or handle. Pilots should further be instructed that if checklist procedures are not effective, they should land at the nearest suitable airport. (A-02-36)

Direct all certificate management offices to instruct inspectors to conduct surveillance of airline dispatch and maintenance control personnel to ensure that their training and operations directives provide appropriate dispatch support to pilots who are experiencing a malfunction threatening safety of flight and instruct them to refrain from suggesting continued flight in the interest of airline flight scheduling. (A-02-37)

As part of the response to Safety Recommendation A-01-41, require operators of Douglas DC-9, McDonnell Douglas MD-80/90, and Boeing 717 series airplanes to remove degraded grease from the jackscrew assembly acme screw and flush degraded grease and particulates from the acme nut before applying fresh grease. (A-02-38)

As part of the response to Safety Recommendation A-01-41, require operators of Douglas DC-9, McDonnell Douglas MD-80/90, and Boeing 717 series airplanes, in coordination with Boeing, to increase the size of the access panels that are used to accomplish the jackscrew assembly lubrication procedure. (A-02-39)

Establish the jackscrew assembly lubrication procedure as a required inspection item that must have an inspector's signoff before the task can be considered complete. (A-02-40)

Review all existing maintenance intervals for tasks that could affect critical aircraft components and identify those that have been extended without adequate engineering justification in the form of technical data and analysis demonstrating that the extended interval will not present any increased risk and require modifications of those intervals to ensure that they (1) take into account assumptions made by the original designers, (2) are supported by adequate technical data and analysis, and (3) include an appropriate safety margin that takes into account the possibility of missed or inadequate accomplishment of the maintenance task. In conducting this review, the Federal Aviation Administration should also consider original intervals recommended or established for new aircraft models that are derivatives of earlier models and, if the aircraft component and the task are substantially the same and the recommended interval for the new model is greater than that recommended for the earlier model, treat such original intervals for the derivative model as "extended" intervals. (A-02-41)

Conduct a systematic industrywide evaluation and issue a report on the process by which manufacturers recommend and airlines establish and revise maintenance task intervals and make changes to the process to ensure that, in the future, intervals for each task (1) take into account assumptions made by the original designers, (2) are supported by adequate technical data and analysis, and (3) include an appropriate safety margin that takes into account the possibility of missed or inadequate accomplishment of the maintenance task. (A-02-42)

Require operators to supply the Federal Aviation Administration (FAA), before the implementation of any changes in maintenance tasks intervals that could affect critical aircraft components, technical data and analysis for each task demonstrating that none of the proposed changes will present any potential hazards, and obtain written approval of the proposed changes from the principal maintenance inspector and written concurrence from the appropriate FAA aircraft certification office. (A-02-43)

Pending the incorporation of a fail-safe mechanism in the design of the Douglas DC-9, McDonnell Douglas MD-80/90, and Boeing 717 horizontal stabilizer jackscrew assembly, as recommended in Safety Recommendation A-02-49, establish an end play check interval
that (1) accounts for the possibility of higher-than-expected wear rates and measurement error in estimating acme nut thread wear and (2) provides for at least two opportunities to detect excessive wear before a potentially catastrophic wear condition becomes possible. (A-02-44)

Require operators to permanently (1) track end play measures according to airplane registration number and jackscrew assembly serial number, (2) calculate and record average wear rates for each airplane based on end play measurements and flight times, and (3) develop and implement a program to analyze these data to identify and determine the cause of excessive or unexpected wear rates, trends, or anomalies. The Federal Aviation Administration (FAA) should also require operators to report this information to the FAA for use in determining and evaluating an appropriate end play check interval. (A-02-45)

Require that maintenance facilities that overhaul jackscrew assemblies record and inform customers of an overhauled jackscrew assembly's end play measurement. (A-02-46)

Require operators to measure and record the on-wing end play measurement whenever a jackscrew assembly is replaced. (A-02-47)

Require that maintenance facilities that overhaul Douglas DC-9, McDonnell Douglas MD-80/90, and Boeing 717 series airplanes' jackscrew assemblies obtain specific authorization to perform such overhauls, predicated on demonstrating that they possess the necessary capability, documentation, and equipment for the task and that they have procedures in place to (1) perform and document the detailed steps that must be followed to properly accomplish the end play check procedure and lubrication of the jackscrew assembly, including specification of appropriate tools and grease types; (2) perform and document the appropriate steps for verifying that the proper acme screw thread surface finish has been applied; and (3) ensure that appropriate packing procedures are followed for all returned overhauled jackscrew assemblies, regardless of whether the assembly has been designated for storage or shipping. (A-02-48)

Conduct a systematic engineering review to (1) identify means to eliminate the catastrophic effects of total acme nut thread failure in the horizontal stabilizer trim system jackscrew assembly in Douglas DC-9 (DC-9), McDonnel Douglas MD-80/90 (MD-80/90), and Boeing 717 (717) series airplanes and require, if practicable, that such fail-safe mechanisms be incorporated in the design of all existing and future DC-9, MD-80/90, and 717 series airplanes and their derivatives; (2) evaluate the horizontal stabilizer trim systems of all other transport-category airplanes to identify any designs that have a catastrophic single-point failure mode and, for any such system; (3) identify means to eliminate the catastrophic effects of that single-point failure mode and, if practicable, require that such fail-safe mechanisms be incorporated in the design of all existing and future airplanes that are equipped with such horizontal stabilizer trim systems (A-02-49)

Modify the certification regulations, policies, or procedures to ensure that new horizontal stabilizer trim control system designs are not certified if they have a single-point catastrophic failure mode, regardless of whether any element of that system is considered structure rather than system or is otherwise considered exempt from certification standards for systems. (A-02-50)

Review and revise aircraft certification regulations and associated guidance applicable to the certification of transport-category airplanes to ensure that wear-related failures are fully considered and addressed so that, to the maximum extent possible, they will not be catastrophic. (A-02-51)

4.2 Previously Issued Recommendations Resulting From This Accident Investigation

As a result of the Alaska Airlines flight 261 accident investigation, the Safety Board issued the following safety recommendations to the FAA on October 1, 2001:

Require the Boeing Commercial Airplane Group to revise the lubrication procedure for the horizontal stabilizer trim system of Douglas DC-9, McDonnell Douglas MD-80/90, and Boeing 717 series airplanes to minimize the probability of inadequate lubrication. (A-01-41)

On June 14, 2002, the Safety Board classified Safety Recommendation A-01-41 "Open—Acceptable Response."

Require the Boeing Commercial Airplane Group to revise the end play check procedure for the horizontal stabilizer trim system of Douglas DC-9, McDonnell Douglas MD-80/90, and Boeing 717 series airplanes to minimize the probability of measurement error and conduct a study to empirically validate the revised procedure against an appropriate physical standard of actual acme screw and acme nut wear. This study should also establish that the procedure produces a measurement that is reliable when conducted on-wing. (A-01-42)

On June 14, 2002, the Safety Board classified Safety Recommendation A-01-42 "Open—Acceptable Response."

Require maintenance personnel who lubricate the horizontal stabilizer trim system of Douglas DC-9, McDonnell Douglas MD-80/90, and Boeing 717 series airplanes to undergo specialized training for this task. (A-01-43)

On June 14, 2002, the Safety Board classified Safety Recommendation A-01-43 "Open—Unacceptable Response."

Require maintenance personnel who inspect the horizontal stabilizer trim system of Douglas DC-9, McDonnell Douglas MD-80/90, and Boeing 717 series airplanes to undergo specialized training for this task. This training should include familiarization with the selection, inspection, and proper use of the tooling to perform the end play check. (A-01-44)

On June 14, 2002, the Safety Board classified Safety Recommendation A-01-44 "Open—Unacceptable Response."

Before the implementation of any proposed changes in allowable lubrication applications for critical aircraft systems, require operators to supply to the FAA technical data (including performance information and test results) demonstrating that the proposed changes will not present any potential hazards and obtain approval of the proposed changes from the principal maintenance inspector and concurrence from the FAA applicable aircraft certification office. (A-01-45)

On July 29, 2002, the Safety Board classified Safety Recommendation A-01-45 "Closed—Acceptable Action."

Issue guidance to principal maintenance inspectors to notify all operators about the potential hazards of using inappropriate grease types and mixing incompatible grease types. (A-01-46)

On July 29, 2002, the Safety Board classified Safety Recommendation A-01-46 "Closed—Acceptable Action."

Survey all operators to identify any lubrication practices that deviate from those specified in the manufacturer's airplane maintenance manual, determine whether any of those deviations involve the current use of inappropriate grease types or incompatible grease mixtures on critical aircraft systems and, if so, eliminate the use of any such inappropriate grease types or incompatible mixtures. (A-01-47)

On June 14, 2002, the Safety Board classified Safety Recommendation A-01-47 "Open—Acceptable Response."

Within the next 120 days, convene an industrywide forum to disseminate information about and discuss issues pertaining to the lubrication of aircraft components, including the qualification, selection, application methods, performance, inspection, testing, and incompatibility of grease types used on aircraft components. (A-01-48)

On June 14, 2002, the Safety Board classified Safety Recommendation A-01-48 "Open—Acceptable Response."

BY THE NATIONAL TRANSPORTATION SAFETY BOARD

CAROL J. CARMODY
Acting Chairman

JOHN J. GOGLIA
Member

JOHN A. HAMMERSCHMIDT
Member

GEORGE W. BLACK, JR.
Member

Adopted: December 30, 2002

Board Member Statements

Acting Chairman Carol J. Carmody's Statement

I believe that NTSB staff presented a strong case in the draft report on Alaska Airlines Flight 261, and during its presentation to the Board at our December 10, 2002 meeting, for recommending that the FAA should convene a headquarters'-led team to conduct another in-depth, on-site follow-up inspection of Alaska Airlines to evaluate whether adequate corrective measures have been fully implemented to cure the deficiencies identified in the FAA's April 2000 special inspection report. I did not support the motion to eliminate this recommendation from the report.

The accident report thoroughly discusses the maintenance failings, and the support for convening a headquarters'-led team can be summarized as follows:

1. The systemic nature and widespread impact of previously identified deficiencies;

2. Alaska Airlines' failure to demonstrate during the FAA's first follow-up inspection (in September 2000) that it had fully corrected the deficiencies;

3. The absence of any in-depth follow-up inspection by an objective team of FAA inspectors not personally responsible for the oversight of Alaska Airlines to verify that the deficiencies have in fact been corrected; and

4. Recent maintenance errors by Alaska Airlines maintenance personnel during end play checks and evidence of continued poor lubrication practices.

The argument made during the meeting that yet another inspection of Alaska Airlines would divert already stretched FAA resources from other important tasks is actually an argument for the FAA to make. The Board's responsibility is to make recommendations to improve safety. In making these recommendations, we try to avoid specific direction on the means to accomplish any given safety objective. I would expect that a recommendation to the FAA to undertake an inspection would be considered as would any other Board recommendation. The FAA would review the need and the resources available and decide whether or not to pursue the recommendation. I would not expect the FAA to pursue an inspection if that pursuit would divert FAA resources from more important safety accomplishments. Moreover, I would expect that if that were a possibility, the FAA would so advise the Board.

There needs to be an assurance that Alaska Airlines has permanently remedied the recurring maintenance problems we have documented in our exhaustive investigation. The public expects and deserves no less. FAA should not have to be coaxed by the NTSB into being more proactive, and I hope that we have not served to perpetuate any problems at Alaska Airlines that the recommended inspection was intended to identify and correct.

Member Black concurred with this statement.

Member John J. Goglia's Statement

This is a maintenance accident. Alaska Airlines' maintenance and inspection of its horizontal stabilizer activation system was poorly conceived and woefully executed. The failure was compounded by poor oversight. Lubrication periods were extended, while inspection intervals were simultaneously lengthened, neither with sound technical basis. And, if logic and standard practice dictate that as risk increases so should monitoring, Alaska's program was otherwise. And in the midst of this, the accident aircraft was dispatched from a C-check with a jackscrew of questionable serviceability that was, in all probability, not greased. And the evidence is that it was never adequately greased again. Had any of the managers, mechanics, inspectors, supervisors, or FAA overseers whose job it was to protect this mechanism done their job conscientiously, this accident cannot happen. The jackscrew is robust. Even at its wear limit condition, it is many times strong enough to carry out its function. And it does not wear quickly when cared for. It is a time-tested mainstay in the fleet, and one major carrier with a diligent approach to its maintenance has never seen one seriously wear, much less wear out. A whole herd of miscues was needed to allow it to fail. Virtually any system on an aircraft treated with the indifference shown to this mechanism will break, many with equally catastrophic effect. Aircraft simply must be maintained, and maintained with care and at all cost.

I say these things because I will also admit that it may be possible to design a better jackscrew. If there is an approach that makes this mechanism fail safe, and if the designers feel confident that the dangers of unintended consequences in changing a proven design are zero or less, then yes, let's by all means add a fail-safe device. But we can easily overdue and over simplify our concern with design issues. We want now to design a safer actuating system and we want to design a better wear measurement system. Fine. But the existing design is safe, it just includes grease as a part of the load transferring system and there wasn't any grease. And the measuring device we now have, no matter how rudimentary it is thought to be, seems to work well for many airlines, and it identified the fact that this jackscrew was wearing and needed to be replaced. The carrier simply failed to do its job.

All the systems in an aircraft that involve large motions under high loads require maintenance. Careful maintenance. And that is a part of the design equation, and once designed, it has to be followed. Certainly we should not rely *solely* on maintenance where mechanical design can compensate for maintenance failure. Obviously, two hands are better than one. Yet it is seductive error to think that we can engineer out failure permanently. We can't. Fail-safe components subject to neglect and abuse are themselves susceptible to failure. So we need to be vigilant in our emphasis on both sides of this equation. This was not a design accident. This was a maintenance accident, perhaps more purely so than any others we have seen. It would be sad and ironic indeed if some new and supposedly fail-safe design were to come to lessen the concentration we put to keeping these mechanisms well-maintained.

Let me reiterate, aircraft must be maintained with care and at all cost. Alaska Airlines expanded rapidly in the years before this accident. With the goal of becoming more profitable, as they became bigger and busier the pressures to keep their planes on schedule put increasing stress on their maintenance facilities. And this took its toll. The Federal Aviation Administration seemed to know this, but was nowhere effective in preventing the tragic chain of events. So N963AS began its fateful path in a C-check years before falling to the ocean. Its maintainers found a jackscrew that needed to be pulled, but no spare was found and, as the part was arguably acceptable, they pushed the plane back into service, with no watch list, no trailers, or orders to keep track of its condition. There were no specific procedures to do so, and no one thought enough to ask for one. The aircraft was arguably, that is technically, legal, and it was probably safe, if it was carefully greased. It was not, we know that without question. And the lengthened inspection intervals were such that it was not to be looked at again, until it was in our laboratory. When it finally failed, ground support from Alaska Airlines seemed to encourage the crew to proceed with a broken plane on to their scheduled destination, for reasons perhaps of convenience both to passengers and maintenance – we won't really ever know. But the impression is inescapable. An aircraft that had been hustled out the door three years earlier for the convenience of scheduling was now encouraged to keep to its appointed routing. It is less coincidence than culture.

NTSB has made several specific maintenance recommendations, some already accomplished, that will, if followed, prevent the recurrence of this particular accident. But maintenance, poorly done, will find a way to bite somewhere else. Alaska needs to re-constitute its will to performance and perfection on the shop floor. FAA needs to revitalize its Alaska staff and become more intelligent about its efforts to oversee. And if, while all this goes on, the manufacturers can also add greater margins for error, that is icing on the cake. I am interested to see what system enhancements come from this, but I am still left with a mechanic's perspective -- you either maintain it or it breaks. This is universally applicable. Like the old adage says, "you schedule maintenance, or the maintenance will schedule you."

Acting Chairman Carmody and Members Hammerschmidt and Black concurred with this statement.

this page intentionally left blank

5. Appendixes

Appendix A
Investigation and Public Hearing

Investigation

The National Transportation Safety Board was initially notified of this accident about 1945 eastern standard time on January 31, 2000. Investigators from the Safety Board's regional office in Los Angeles, California, went immediately to the scene of the accident. A full go-team was assembled in Washington, D.C., and arrived on scene early the next morning. The go-team was accompanied by Member John Hammerschmidt and representatives from the Board's Office of Government, Public, and Family Affairs.

Parties to the investigation were the Federal Aviation Administration (FAA); the Boeing Commercial Airplane Group; Alaska Airlines, Inc.; the Aircraft Mechanics Fraternal Association (AMFA); the Air Line Pilots Association (ALPA); the Association of Flight Attendants; the National Air Traffic Controllers Association; Pratt & Whitney, and Shell Global Solutions.

Public Hearing

A public hearing was conducted for this accident from December 13 to 15, 2000, in Washington, D.C. Member John Hammerschmidt presided over the hearing. Parties to the public hearing were the FAA, Alaska Airlines, Boeing, ALPA, and AMFA.

Appendix B

Appendix B
Cockpit Voice Recorder Transcript

The following is a transcript of the Fairchild A-100A cockpit voice recorder (CVR) serial number 62892, installed on the accident airplane. The CVR transcript reflects the 31 minutes before power was lost to the CVR. All times are Pacific standard time, based on a 24-hour clock.

LEGEND

RDO	Radio transmission from accident aircraft, Alaska 261
CAM	Cockpit area microphone voice or sound source
PA	Voice or sound heard on the public address system channel.
HOT	Hot microphone voice or sound source[1]

For RDO, CAM, HOT, and PA comments:
- -1   Voice identified as the Captain
- -2   Voice identified as the First Officer
- -3   Voice identified as a Flight Attendant
- -?   Voice unidentified

MZT	Radio transmission from Mazetlan Center
LAX CTR1	Radio transmission from the Los Angeles Air Route Traffic Control Center sector 30 controller
LAX CTR2	Radio transmission from the Los Angeles Air Route Traffic Control Center sector 25 controller
LAX-MX	Radio transmission from Alaska Airlines Maintenance facility in Los Angeles
LAX-OPS	Radio transmission from Alaska Airlines Operations facility in Los Angeles
SEA-DIS	Radio transmission from Alaska Airlines Dispatch facility in Seattle
SEA-MX	Radio transmission from Alaska Airlines maintenance facility in Seattle

- -1   First voice
- -2   Second voice

---

[1] This recording contained some audio from one Hot microphone. The voices or sounds heard on this channel were also heard on the CAM channel, and are annotated as coming from the CAM channel in this transcript. The audio from the HOT microphone was used to clarify the CAM audio when possible.

ATIS	Radio transmission from Los Angeles airport Automated Terminal Information System
CAWS	Mechanical voice or sound source from the Central Aural Warning System, as heard on the Cockpit Area Microphone channel.
*	Unintelligible word
@	Non-pertinent word
#	Expletive
- - -	Break in continuity or interruption in comment
( )	Questionable insertion
[ ]	Editorial insertion
...	Pause

Note 1: Times are expressed in Pacific Standard Time (PST).

Note 2: Generally, only radio transmissions to and from the accident aircraft were transcribed.

Note 3: Words shown with excess vowels, letters, or drawn out syllables are a phonetic representation of the words as spoken.

Note 4: A non pertinent word, where noted, refers to a word not directly related to the operation, control or condition of the aircraft.

Appendix B 194 **Aircraft Accident Report**

INTRA-COCKPIT COMMUNICATION		AIR-GROUND COMMUNICATION	
TIME and SOURCE	**CONTENT**	**TIME and SOURCE**	**CONTENT**
Start of Transcript			
1549:49.3 [start of recording]			
		1549:50 SEA-MX	um beyond that I have verified no history on your aircraft in the past thirty days.
1549:54.8 CAM	[sound of click]		
		1549:57.7 RDO-1	yea we didn't see anything in the logbook.
1550:14 CAM-2	why don't you pull your your seat forward and I'll just check this pedestal back there. I don't think there's anything beyond that we haven't checked.		
1550:22 CAM-1	see when he's saying pedestal... I believe he's talking about this---		
1550:25 CAM-2	oh.		
1550:25 CAM-1	---switch that's on the * that's on the pedestal.		
1550:27 CAM-2	yea okay.		
1550:31 CAM-1	do you see anything back there?		

Appendix B — 195 — Aircraft Accident Report

INTRA-COCKPIT COMMUNICATION

TIME and SOURCE	CONTENT
1550:32 CAM	[Sound of click]
1550:33 CAM-2	uh there's *.

AIR-GROUND COMMUNICATION

TIME and SOURCE	CONTENT
1550:40 SEA-MX	and two sixty one, maintenance.
1550:42.0 RDO-1	go ahead maintenance two six one.
1550:44 SEA-MX	understand you're requesting uh diversion to L A for this uh discrepancy is there a specific reason you prefer L A over San Francisco?
1550:45 MZT	Alaska two sixty one radar service terminated.... contact uh Los Angeles center frequency one one nine decimal ninety five good day.
1550:54.4 RDO-1	well a lotta times its windy and rainy and wet in San Francisco and uh, it seemed to me that a dry runway... where the wind is usually right down the runway seemed a little more reasonable.
1550:55.0 RDO-2	one one nine ninety five Alaska two sixty one.
1551:01.2 RDO-2	say again the frequency one one nine point eh ninety five?

Appendix B 196 **Aircraft Accident Report**

INTRA-COCKPIT COMMUNICATION		AIR-GROUND COMMUNICATION	
TIME and SOURCE	CONTENT	TIME and SOURCE	CONTENT
		1551:05 MZT	affirm one one nine decimal ninety five.
		1551:09.3 RDO-2	roger.
		1551:09.9 SEA-MX	ok and uh.... is this added fuel that you're gonna have in L A gonna be a complication or an advantage?
		1551:18.1 RDO-1	well the way I'm reading it uh heavier airplanes land faster.... right now I got fifteen five on board, I'm thinking to land with about twelve which is still uh an hour and forty minutes.... uh and those are the numbers I'm running up here.
		1551:20.6 RDO-2	L A Alaska two sixty one three one zero.
		1551:36 SEA-MX	ok uh two sixty one standby for dispatch.
		1551:38 RDO-2	Los Angeles Alaska two sixty one three one zero.
		1551:40 RDO-1	OK the other thing you gotta know is that they're talking about holding and delays in San Francisco um for your maintenance facil- eh you know planning uh it uh L A seemed like a smarter move from airworthy move.
		1551:42 LAX-CTR1	Alaska two sixty one L A center roger.

Appendix B   197   Aircraft Accident Report

INTRA-COCKPIT COMMUNICATION

TIME and SOURCE	CONTENT

AIR-GROUND COMMUNICATION

TIME and SOURCE	CONTENT
1551:50 RDO-2	* there's two people on the frequency I'm sorry Alaska two sixty one I didn't hear your response.
1551:58 LAX-CTR1	Alaska two six one squawk two zero one zero.
1552:01 RDO-2	two zero one zero Alaska two sixty one.
1552:02 SEA-DIS	two sixty one dispatch… uh current San Francisco weather one eight zero at six, nine miles, few at fifteen hundred broken twenty eight hundred overcast thirty four hundred…. uh if uh you want to land at L A of course for safety reasons we will do that uh wu we'll uh tell you though that if we land in L A uh we'll be looking at probably an hour to an hour and a half we have a major flow program going right now. uh that's for ATC back in San Francisco.
1552:31 RDO-1	well uh yu you eh huh… boy you put me in a spot here um…..
1552:41 RDO-1	I really didn't want to hear about the flow being the reason you're calling us cause I'm concerned about overflying suitable airports.
1552:51 SEA-DIS	well we we wanna do what's safe so if that's what you feel is uh safe we just wanna make sure you have all of the uh…. all the info.

Appendix B 198 **Aircraft Accident Report**

INTRA-COCKPIT COMMUNICATION		AIR-GROUND COMMUNICATION	
TIME and SOURCE	**CONTENT**	**TIME and SOURCE**	**CONTENT**
		1552:59 RDO-1	yea we we kinda assumed that we had... what's the uh the wind again there in San Francisco?
		1553:03 SEA-DIS	wind at San Francisco currently zero uh one zero eight at six.
1553:08 CAM-2	what runway they landing... one zero?		
1553:09 CAM-1	what's that?		
1553:10 CAM-2	ask him what runway they're landing.		
		1553:11 RDO-1	and confirm they're landing runway one zero?
		1553:15 SEA-DIS	and uh standby I'll confirm that.
1553:17 CAM-2	and see if the runways are dry or wet.		
		1553:19 RDO-1	and we need to know if they're dry or wet.
		1553:21 SEA-DIS	eh yup I'll uh find that out and uh correction on that wind one eight zero at six and standby.

INTRA-COCKPIT COMMUNICATION		AIR-GROUND COMMUNICATION	
TIME and SOURCE	CONTENT	TIME and SOURCE	CONTENT
1553:28 CAM-1	one eight zero at six... so that's runway one one six what we need is runway one nine, and they're not landing runway one nine.		
1553:35 CAM-2	I don't think so.		
1553:37 CAM-2	we might just ask if there's a ground school instructor there available and and discuss it with him... or a uh simulator in-structor.		
1553:40 CAM-1	yea.		
		1553:46 RDO-1	and uh dispatch one sixty one... we're wondering if we can get some support out of the uh instructernal force---
		1553:53 RDO-1	---instructors up there if they got any ideas on us.
1554:23 CAM-1	you're talkin to ATC huh?		
1554:24 CAM-2	yea uh huh.		
1554:26 CAM-2	well lets confirm the route of flight its uh, I wasn't totally sure but its uh direct Oceanside?		

Appendix B 200 **Aircraft Accident Report**

INTRA-COCKPIT COMMUNICATION		AIR-GROUND COMMUNICATION
TIME and SOURCE	**CONTENT**	**TIME and SOURCE**	**CONTENT**

TIME and SOURCE	CONTENT
1554:32 CAM-1	Tijuana Oceanside... Oceanside right... then Santa Catalina.
1554:47 CAM-1	ehh somebody was callin in about wheelchairs---
1554:50 CAM-3	oh really?
1554:50 CAM-1	---when I'm workin a problem.
1554:51 CAM-3	is that why it went static?
1554:53 CAM-1	ok yea now... I just that's something that oughta be in the computers, if they want it that bad they you guys oughta be able to pick up the phone---
1555:00 CAM-3	mmm hmm.
1555:00 CAM-1	---just... drives me nuts. not that I wanna go on about it... you know I it just blows me away they think we're gonna land, they're gonna fix it, now they're worried about the flow, I'm sorry this airplane's idn't gonna go anywhere for a while.... so you know.
1555:16 CAM-3	so they're trying to put the pressure on you---

Appendix B 201 **Aircraft Accident Report**

INTRA-COCKPIT COMMUNICATION

TIME and SOURCE	CONTENT
1555:18 CAM-1	well no, yea.
1555:19 CAM-3	---well get it to where it needs to be.
1555:20 CAM-1	and actually it doesn't matter that much to us.
1555:23 CAM-3	still not gonna go out on time to the next *.
1555:24 CAM-1	yea... yea... I thought they'd cover the people better from L A---
1555:29 CAM-3	L A
1555:30 CAM-1	---then San Francisco.

AIR-GROUND COMMUNICATION

TIME and SOURCE	CONTENT
1555:32 RDO-2	L A Alaska two sixty one just confirm our routing after uh Tijuana is, direct Oceanside?
1555:38 LAX-CTR1	Alaska two sixty one after Tijuana cleared to San Francisco via direct San Marcos jet five zero one Big Sur direct maintain flight level three one zero.
1555:47 RDO-2	OK uh San Francisco San Marcos J five zero one Big Sur uh direct three one zero Alaska two sixty one.

Appendix B — 202 — Aircraft Accident Report

INTRA-COCKPIT COMMUNICATION

TIME and SOURCE	CONTENT

AIR-GROUND COMMUNICATION

TIME and SOURCE	CONTENT
1555:55 LAX-CTR1	*
1556:03 SEA-DIS	Alaska two sixty one dispatch.
1556:06 RDO-1	dispatch Alaska two six one go ahead.
1556:08 SEA-DIS	yea I called uh ATIS they're landing two eight right two eight left and uh wasn't able to get the the runway report but uh looking at past uh weather it hasn't rained there in hours so I'm looking at uh probably a dry runway.
1556:21 RDO-1	ok uh.
1556:26 RDO-1	I have with the information I have available to me and we're waitin on that CG update I'm looking at a uh approach speed of a hundred and eighty knots, uh do you have a wind at L A X lax?
1556:50 SEA-DIS	its two six zero at nine.
1556:56 RDO-1	ok two six at nine.....
1556:59 RDO-1	...versus a direct crosswind which is effectively no change in groundspeed... I gotta tell you, when I look at it from a safety point I think that something that lowers my ground-speed makes sense.

Appendix B — 203 — Aircraft Accident Report

	INTRA-COCKPIT COMMUNICATION	AIR-GROUND COMMUNICATION	
TIME and SOURCE	CONTENT	TIME and SOURCE	CONTENT
		1557:16 SEA-DIS	ok two sixty one that'll uh that'll mean L A X then for you um I was gonna get you if I could to call L A X with that uh info and they can probably whip out that CG for you real quick.
		1557:30 RDO-1	I suspect that uh that's what we'll have to do. ok here's uh, my plan is we're gonna continue as if going to San Francisco get all that data then begin our descent back in to L A X, and at a lower altitude we will configure, and check the handling uh envelope before we proceed with the approach.
		1558:05 SEA-DIS	ok two sixty one dispatch copied that, if you can now keep uh L A ops updated on uh your ETA, that would be great and I'll be talking with them.
		1558:15 RDO-1	ok well ah if you'll let them know we're comin here I'll I think they'll probably listen as we talk.... were goin to L A X were gonna stay up here and burn a little more gas get all our ducks in a row, and then we'll uh be talking to L A X when we start down to go in there.
		1558:29 SEA-DIS	ok and if you have any problems with them giving you a CG gimme a call back.
		1558:34 RDO-1	ok. break, L A X do you read Alaska two six one?
		1558:39 LAX-OPS	two sixty one I do copy do you have an ETA for me?

Appendix B **Aircraft Accident Report**

AIR-GROUND COMMUNICATION

TIME and SOURCE	CONTENT
1558:43 RDO-1	well....
1558:45 RDO-1	...yea I'm gonna put it at about thirty, thirty five minutes, I could actually, the longer the more fuel I burn off the better I am.... but I wonder if you can compute our current CG based on the information we had at takeoff for me.
1558:58 LAX-OPS	ok you're transmission is coming in broken but uh, go ahead.
1559:02 RDO-1	you know what I'll wait a minute we'll be a little bit closer and that'll help everything.
1559:06 LAX-OPS	ok also uh two sixty one just be advised uh because you're an international arrival we have to get landing rights I don't know how long that's gonna take me... but uh I have to clear it all through customs first.
1559:19 RDO-1	ok I unders... I remember this is complicated, yea well, better start that now cause we are comin to you.
1559:26 LAX-OPS	copy.

INTRA-COCKPIT COMMUNICATION

TIME and SOURCE	CONTENT
1559:29 CAM-1	we'll call em back over as we get closer to Catalina.
1559:34 CAM-2	as we get what?

Appendix B   205   Aircraft Accident Report

INTRA-COCKPIT COMMUNICATION

TIME and SOURCE	CONTENT
1559:34 CAM-1	closer to L A she's got to get landing rights.
1559:37 CAM-2	were ninety four miles from L A now.
1559:38 CAM-1	oh ok. you wanna listen to the ATIS you can.
1559:42 CAM-2	in fact I switched it once already just kinda late.
1559:44 CAM-1	you got the jet.
1559:44 CAM-2	I got it.

AIR-GROUND COMMUNICATION

TIME and SOURCE	CONTENT

Appendix B 206 **Aircraft Accident Report**

INTRA-COCKPIT COMMUNICATION

TIME and SOURCE	CONTENT
1601:01 CAM-2	so he wanted us to go to San Fran initially?
1601:06 CAM-1	to keep the schedule alive. I mean it was just... it was I mean he had all the reasons to do it, I stated concern about flying overflying a suitable airport—
1601:15 CAM-2	yea.
1601:16 CAM-1	—but I was listening, then when he gives me the wind, its it's... the wind was a ninety degree cross at ten knots. two eight and we'd be landing on—

AIR-GROUND COMMUNICATION

TIME and SOURCE	CONTENT
1559:50 ATIS	charlie five and charlie six is restricted * taxiway charlie five is restricted to MD eleven and smaller. read back all runway hold short instructions. upon receipt of your ATC clearance read back only your callsign and transponder code unless you have a question. advise on initial contact, you have information mike. Los Angeles international airport information mike. two two five zero zulu. wind two three zero at eight. visibility eight. few clouds at two thousand eight hundred. one two thousand scattered. ceiling two zero thousand overcast. temperature one six dewpoint one one. altimeter three zero one seven. simultaneous ILS approaches in progress runway two four right and two five left or vector for visual approach will be provided. simultaneous visual approaches to all runways are in progress. and parallel localizer approaches are in progress between Los Angeles international and Hawthorne airports. simultaneous instrument departure in progress runway two four and two five. notices to airmen.

Appendix B 207 **Aircraft Accident Report**

INTRA-COCKPIT COMMUNICATION

TIME and SOURCE	CONTENT
1601:30 CAM-2	and they are using one nine?
1601:33 CAM-1	you know I don't know... I wrote it down there... the winds were... one eighty at six.... I don't know.
1601:49 CAM-2	I don't know.
1601:49 CAM-1	I don't care... you know what? I expect him to figure all that # ---
1601:53 CAM-2	right.
1601:53 CAM-1	---he's got it on the screen---
1601:54 CAM-2	that's why I was thinking that an instructor would really uh---
1601:58 CAM-1	yea.
1601:58 CAM-2	---cut through the crap there.
1601:59 CAM-2	they... not available?
1602:00 CAM-1	well they just don't talk to each other.

AIR-GROUND COMMUNICATION

TIME and SOURCE	CONTENT

Appendix B 208 **Aircraft Accident Report**

INTRA-COCKPIT COMMUNICATION		AIR-GROUND COMMUNICATION	
TIME and SOURCE	**CONTENT**	**TIME and SOURCE**	**CONTENT**
1602:02 CAM-2	oh.		
1602:02 CAM-1	I mean I * ----		
1602:04 CAM-2	* they've always told us they were available you know---		
1602:06 CAM-1	yea yea.		
1602:07 CAM-2	----anytime you have a problem.		
1602:09 CAM-2	if they get one down there.		
		1602:12.6 RDO-1	Los Angeles one sixty one do you read me better now?
1602:29 CAM-1	I got the track goin over there.		
		1602:31 LAX-OPS	go ahead two six one.
		1602:33.6 RDO-1	two sixty one, I.... I know you're busy on us uh, but we're discussing it up here could you give us the winds at San Francisco if you could just pull em up on your screen?
1602:57 CAM-2	I thought they....		

INTRA-COCKPIT COMMUNICATION		AIR-GROUND COMMUNICATION	
TIME and SOURCE	**CONTENT**	**TIME and SOURCE**	**CONTENT**
		1603:00 LAX-OPS	ok ahhh San Francisco, ok we've got uh... winds are one seventy at six knots.
		1603:15.6 RDO-1	ok thank you that's what I needed. we are comin in to see you.... and I've misplaced the paperwork here.
1603:23 CAM-2	there it is.		
1603:35 CAM-1	I can't read your writing... can you read her the uh zero fuel weight----		
1603:40 CAM-2	yea.		
1603:41 CAM-1	----and all those numbers and CG.		
		1603:43 LAX-OPS	L A operations from two six to two six one.
1603:48 CAM-2	I got it.		
		1603:48.5 RDO-2	uhhh two sixty one... do you need our uh, our numbers?
		1603:52 LAX-OPS	yea we just wanna advise that we do not have landing rights as yet.

INTRA-COCKPIT COMMUNICATION		AIR-GROUND COMMUNICATION	
TIME and SOURCE	**CONTENT**	**TIME and SOURCE**	**CONTENT**
		1603:56 RDO-2	here's our numbers we had uh ten in first class, seventy in coach, zero fuel weight one zero two one one zero point one fuel on board thirty four point niner take off weight one thirty six five one one point eight, CG eleven point eight.
		1604:19 LAX-OPS	OK I got ten and seventy Z fuel weight one zero two one one zero point one, fuel on board thirty four decimal nine take off weight five one one decimal eight and a CG of eleven decimal eight.
		1604:32 RDO-2	yea uh take off one three six five one one point eight and uh CG one one point eight. and we currently have thirteen thousand six hundred pounds of fuel on board.
1604:43 CAM-1	estimate ten thousand on landing.	1604:45 RDO-2	estimating ten thousand pounds on landing.
		1604:53 LAX-OPS	ok you said your takeoff weight was... one one uhh one five one one decimal eight?
		1604:58 RDO-2	one three six five one one point eight.
		1605:05 LAX-OPS	one three six five one one point eight thank you.

INTRA-COCKPIT COMMUNICATION		AIR-GROUND COMMUNICATION	
TIME and SOURCE	**CONTENT**	**TIME and SOURCE**	**CONTENT**
		1605:07 RDO-2	and we're currently a hundred and fifteen seven on our weight, and we'll burn another three thousand pounds.
1605:19 CAM	[sound of two clicks]		
1605:27 CAM-2	I'm back on the uh I'm off of the uh company.		
1606:26 CAM-1	no... that's what I was expecting them to do. duh.		
1606:47 CAM-2	so our... actually our landing speed will be one forty eight plus... some additive right?		
1607:06 CAM-1	lets guess... lets guess one twelve.		
1607:10 CAM-2	ok.		
1607:10 CAM-1	one forty six... plus... I get a minus two, worst case... twenty four knots... fifty sixty seventy... * .		
		1607:33 LAX-OPS	Alaska two sixty one from operations can you give us your tail number?
		1607:38 RDO-1	uh two sixty one, it was ship number nine six three.

Appendix B 212 **Aircraft Accident Report**

INTRA-COCKPIT COMMUNICATION

TIME and SOURCE	CONTENT

AIR-GROUND COMMUNICATION

TIME and SOURCE	CONTENT
1607:43 LAX-OPS	copy that two… uh your aircraft number is nine six three.
1607:47 RDO-1	affirmative thank you.
1607:51 LAX-MX-1	and two sixty one maintenance.
1607:53 RDO-1	two sixty one go.
1607:54 LAX-MX-1	yea are you guys with the uh, horizontal situation?
1607:58 RDO-1	affirmative.
1607:59 LAX-MX-1	yea did you try the suitcase handles and the pickle switches right?
1608:03 RDO-1	yea we tried everything together, uh…..
1608:08 RDO-1	…we've run just about everything if you've got any hidden circuit breakers we'd love to know about 'em.
1608:14 LAX-MX-1	I'm off I'll look at the uh circuit breaker uh guide just as a double check and um yea I just wanted to know if you tried the pickle switches and the suitcase handles to see if it was movin in with any of the uh other switches other than the uh suitcase handles alone or nothing.

Appendix B 213 **Aircraft Accident Report**

INTRA-COCKPIT COMMUNICATION		AIR-GROUND COMMUNICATION	
TIME and SOURCE	**CONTENT**	**TIME and SOURCE**	**CONTENT**
		1608:29.9 RDO-1	yea we tried just about every iteration.
		1608:32 LAX-MX-1	and alternate's inop too huh?
		1608:35.1 RDO-1	yup its just it appears to be jammed the uh the whole thing, it spikes out when we use the primary, we get AC load that tells me the motor's tryin to run but the brake won't move it. when we use the alternate, nothing happens.
		1608:50 LAX-MX-1	ok and you you say you get a spike when on the meter up there in the cockpit when you uh try to move it with the uh um with the primary right?
1608:59 CAM-1	I'm gonna click it off you got it.		
1609:00 CAM-2	ok.		
		1609:01.5 RDO-1	affirmative we get a spike when we do the primary trim but there's no appreciable uh change in the uh electrical uh when we do the alternate.
		1609:09 LAX-MX-1	ok thank you sir see you here.
		1609:11 RDO-1	ok.

Appendix B 214 **Aircraft Accident Report**

INTRA-COCKPIT COMMUNICATION		AIR-GROUND COMMUNICATION	
TIME and SOURCE	**CONTENT**	**TIME and SOURCE**	**CONTENT**
1609:13 CAM-1	lets do that.		
1609:14.8 CAM	[sound of click]		
1609:14.8 CAM-1	this'll click it off.		
1609:16 CAM	[sound of clunk]		
1609:16.9 CAM	[sound of two faint thumps in short succession]		
1609:17.0 CAWS	[sound similar to horizontal stabilizer-in-motion audible tone]		
1609:18 CAM-1	holy #.		
1609:19.6 CAWS	[sound similar to horizontal stabilizer- in-motion audible tone]		
1609:21 CAM-1	you got it?... # me.		
1609:24 CAM-2	what are you doin?		
1609:25 CAM-1	I it clicked off—		

Appendix B 215 **Aircraft Accident Report**

INTRA-COCKPIT COMMUNICATION

TIME and SOURCE	CONTENT
1609:25.4 CAWS	[sound of chime] Altitude
1609:26 CAM-1	---it * got worse.... ok.
1609:30 CAM	[sound similar to airframe vibration begins]
1609:31 CAM-1	you're stalled.
1609:32 CAM	[sound similar to airframe vibration becomes louder]
1609:33 CAM-1	no no you gotta release it ya gotta release it.
1609:34 CAM	[sound of click]
1609:34 CAM	[sound similar to airframe vibration ends]
1609:42.4 CAM-1	lets * speedbrake.
1609:46 CAM-1	gimme a high pressure pumps.
1609:52 CAM-2	ok.
1609:52 CAM-1	help me back help me back.

AIR-GROUND COMMUNICATION

TIME and SOURCE	CONTENT

INTRA-COCKPIT COMMUNICATION		AIR-GROUND COMMUNICATION	
TIME and SOURCE	**CONTENT**	**TIME and SOURCE**	**CONTENT**
1609:54 CAM-2	ok.		
		1609:55 RDO-1	center Alaska two sixty one we are uh in a dive here.
		1610:01.6 RDO-1	and I've lost control, vertical pitch.
1610:01.9 CAWS	[sound of clacker] Overspeed. (begins and repeats for approx 33 seconds)		
		1610:05 LAX-CTR1	Alaska two sixty one say again sir.
		1610:06.6 RDO-1	yea were out of twenty six thousand feet, we are in a vertical dive.... not a dive yet... but uh we've lost vertical control of our airplane.
1610:15 CAM	[sound of click]		
1610:20 CAM-1	just help me.		
1610:22 CAM-1	once we get the speed slowed maybe... we'll be ok.		
		1610:28.2 RDO-1	we're at twenty three seven request uh.
		1610:33 RDO-1	yea we got it back under control here.

INTRA-COCKPIT COMMUNICATION		AIR-GROUND COMMUNICATION	
TIME and SOURCE	**CONTENT**	**TIME and SOURCE**	**CONTENT**
1610:37 CAM-1	ok.	1610:34 RDO-2	no we don't, ok.
1610:40 CAM	[sound of click]	1610:37 LAX-CTR1	the altitude you'd like to uh to remain at?
1610:45 CAM-2	lets take the speedbrakes off I'm * ---		
1610:46 CAM-1	no no leave them there. it seems to be helping.		
1610:51 CAM-1	# me.		
1610:53 CAWS	[sound of chime] Altitude		
1610:55 CAM-1	ok it really wants to pitch down.		
1610:58 CAM-2	ok.		
1610:59 CAM-1	don't mess with that.		
1611:04 CAM-2	I agree with you.		

Appendix B 218 **Aircraft Accident Report**

AIR-GROUND COMMUNICATION

TIME and SOURCE		CONTENT
1611:04	LAX-CTR1	Alaska two sixty one say your condition.
1611:06.6	RDO-1	two sixty one we are at twenty four thousand feet, kinda stabilized.
1611:10	RDO-1	we're slowing here, and uh, we're gonna uh.
1611:15	RDO-1	do a little troubleshooting, can you gimme a block between uh, twenty and twenty five?
1611:21	LAX-CTR1	Alaska two sixty one maintain block altitude flight level two zero zero through flight level two five zero.
1611:27	RDO-1	Alaska two sixty one we'll take that block we'll be monitor'n the freq.

INTRA-COCKPIT COMMUNICATION

TIME and SOURCE		CONTENT
1611:31	CAM-2	you have the airplane let me just try it.
1611:33	CAM-1	ok.
1611:33	CAM-2	uh how hard is it?
1611:33	CAM-1	I don't know my adrenaline's goin.... it was really tough there for a while.

INTRA-COCKPIT COMMUNICATION

TIME and SOURCE	CONTENT
1611:38 CAM-2	yea it is.
1611:39 CAM-1	ok.
1611:43 CAM-2	whatever we did is no good, don't do that again.
1611:44 CAM-1	yea, no it went down it went to full nose down.
1611:48 CAM-2	uh it's a lot worse than it was?
1611:50 CAM-1	yea yea we're in much worse shape now.
1611:59 CAM-1	I think its at the stop, full stop... and I'm thinking, we can- can it go any worse... but it probably can... but when we slowed down, lets slow it lets get down to two hundred knots and see what happens.
1612:16 CAM-2	ok?
1612:16 CAM	[sound of click]
1612:17 CAM-2	we have to put the slats out and everything... flaps and slats.

AIR-GROUND COMMUNICATION

TIME and SOURCE	CONTENT

Appendix B 220 **Aircraft Accident Report**

INTRA-COCKPIT COMMUNICATION		AIR-GROUND COMMUNICATION	
TIME and SOURCE	**CONTENT**	**TIME and SOURCE**	**CONTENT**
1612:20 CAM-1	yea... well we'll wait ok you got it for a second?		
1612:23 CAM-2	yea.		
		1612:25.3 RDO-1	maintenance two sixty one are you on?
		1612:30 LAX-MX-2	yea two sixty one this is maintenance.
		1612:32.0 RDO-1	ok we did---
		1612:33.2 RDO-1	---we did both the pickle switch and the suitcase handles and it ran away full nose trim down.
		1612:39 LAX-MX-2	oh it ran away trim down.
		1612:42 RDO-1	and now we're in a * pinch so we're holding uh we're worse than we were.
		1612:50 LAX-MX-2	ok uh... geez.
		1612:52 LAX-MX-1	you want me to talk to em? (in the background during previous transmission)

Appendix B 221 **Aircraft Accident Report**

INTRA-COCKPIT COMMUNICATION		AIR-GROUND COMMUNICATION	
TIME and SOURCE	**CONTENT**	**TIME and SOURCE**	**CONTENT**
		1612:55 LAX-MX-1	yea two sixty one maintenance uh uh you getting full nose trim down but are you getting any you don't get no nose trim up is that correct?
		1613:04 RDO-1	that's affirm we went to full nose down and I'm afraid to try it again to see if we can get it to go in the other direction.
		1613:10 LAX-MX-1	ok well your discretion uh if you want to try it, that's ok with me if not that's fine. um we'll see you at the gate.
1613:20 CAM-2	did it happen went in reverse? when you pulled back it went forward?		
1613:22 CAM-1	I went tab down... right, and it should have come back instead it went the other way.		
1613:29 CAM-2	uh huh.		
1613:30 CAM-1	what do you think?		
1613:32 CAM-2	uhhh.		
1613:32 CAM-1	you wanna try it or not?		
1613:32 CAM-2	uhh no. boy I don't know.		

Appendix B 222 **Aircraft Accident Report**

INTRA-COCKPIT COMMUNICATION

TIME and SOURCE	CONTENT
1613:33 CAM-1	its up to you man.
1613:35 CAM-2	lets head back toward uh here lets see… well we're---
1613:39 CAM-1	I like where were goin out over the water myself… I don't like goin this fast though.
1613:50 CAM	[sound of click]
1613:57 CAM-1	ok you got * [sound similar to short interruption in recording] second?
1613:58 CAM-2	yea.
1613:59 CAM-2	we better… talk to the people in the back there.
1614:03 CAM-1	yea I know.

AIR-GROUND COMMUNICATION

TIME and SOURCE	CONTENT
1614:04 LAX-CTR1	Alaska two sixty one let me know if you need anything.
1614:08 RDO-2	yea we're still workin this.

Appendix B 223 **Aircraft Accident Report**

INTRA-COCKPIT COMMUNICATION		AIR-GROUND COMMUNICATION	
TIME and SOURCE	**CONTENT**	**TIME and SOURCE**	**CONTENT**
1614:12 PA-1	folks we have had a flight control problem up front here we're workin it uh that's Los Angeles off to the right there that's where we're intending to go. we're pretty busy up here workin this situation I don't anticipate any big problems once we get a couple of sub systems on the line. but we will be going into L A X and I'd anticipate us parking there in about twenty to thirty minutes.		
1614:39 CAM-1	ok... did the, first of all, speedbrakes. did they have any effect?		
1614:49 CAM-1	lets put the power where it'll be for one point two, for landing. you buy that?		
1614:53 CAM-1	slow it down and see what happens.		
		1614:54 LAX-CTR1	Alaska two sixty one contact L A center one two six point five two they are aware of your situation.
		1615:00.0 RDO-2	ok Alaska two sixty one say again the frequency, one two zero five two?
1615:02 CAM-1	I got the yoke.		
		1615:04 LAX-CTR1	Alaska two sixty one, twenty six fifty two.

Appendix B 224 **Aircraft Accident Report**

	INTRA-COCKPIT COMMUNICATION		AIR-GROUND COMMUNICATION
TIME and SOURCE	**CONTENT**	**TIME and SOURCE**	**CONTENT**
		1615:06 RDO-2	thank you.
		1615:07 LAX-CTR1	you're welcome have a good day.
		1615:19.7 RDO-2	L A Alaska two sixty one we're with you we're at twenty two five, we have a jammed stabilizer and we're maintaining altitude with difficulty. uh but uh we can maintain altitude we think... and our intention is to land at Los Angeles.
		1615:36 LAX-CTR2	Alaska two sixty one L A center roger um you're cleared to Los Angeles airport via present position direct uh Santa Monica, direct Los Angeles and uh, you want lower now or what do you want to do sir?
1615:54 CAM-1	let me get let me have it.		
		1615:56 RDO-1	center uh Alaska two sixty one. I need to get down about ten, change my configuration, make sure I can control the jet and I'd like to do that out here over the bay if I may.
		1616:07 LAX-CTR2	ok Alaska two sixty one roger that standby here.
1616:11 CAM-2	lets do it at this altitude instead—		
1616:11 CAM-1	what?		

Appendix B 225 **Aircraft Accident Report**

INTRA-COCKPIT COMMUNICATION		AIR-GROUND COMMUNICATION	
TIME and SOURCE	**CONTENT**	**TIME and SOURCE**	**CONTENT**
1616:12 CAM-2	----of goin to ten lets do it at this altitude.		
1616:14 CAM-1	cause the airflow's that much difference down at ten this air's thin enough that that you know what I'm sayin?		
1616:20 CAM-2	yea uh I'll tell em to uh----		
1616:22 CAM-1	I just made a PA to everyone to get everybody----		
1616:24 CAM-2	ok.		
1616:26 CAM-1	----down you might call the flight attendants.		
1616:27 CAM	[sound similar to cockpit door operating]		
1616:32 CAM-3	I was just comin up this way.		
		1616:32 LAX-CTR2	Alaska two sixty one fly a heading of two eight zero and descend and maintain one seven thousand.
1616:34 CAM-2	uhh.		
1616:36 CAM	[sound similar to cockpit door operating]		

Appendix B 226 **Aircraft Accident Report**

AIR-GROUND COMMUNICATION

TIME and SOURCE	CONTENT
1616:39.0 RDO-1	two eight zero and one seven seventeen thousand Alaska two sixty one. and we generally need a block altitude.
1616:45 LAX-CTR2	ok and just um I tell you what do that for now sir, and contact L A center on one three five point five they'll have further uhh instructions for you sir.
1616:56.9 RDO-2	ok thirty five five say the altimeter setting?
1616:59 LAX-CTR2	the L A altimeter is three zero one eight.
1617:02 RDO-2	thank you.

INTRA-COCKPIT COMMUNICATION

TIME and SOURCE	CONTENT
1617:01 CAM-1	I need everything picked up---
1617:02 CAM-1	---and everybody strapped down---
1617:04 CAM-3	ok.
1617:04 CAM-1	---cause I'm gonna unload the airplane and see if we can---
1617:06 CAM-3	ok.
1617:07 CAM-1	---we can regain control of it that way.

Appendix B 227 **Aircraft Accident Report**

INTRA-COCKPIT COMMUNICATION		AIR-GROUND COMMUNICATION	
TIME and SOURCE	**CONTENT**	**TIME and SOURCE**	**CONTENT**
1617:09 CAM-3	ok we had like a big bang back there---		
1617:11 CAM-1	yea I heard it---		
1617:12 CAM-3	ok.		
1617:12 CAM-1	---the stab trim I think it---		
1617:13 CAM-2	you heard it in the back?		
1617:13 CAM-3	yea.		
1617:14 CAM-2	yea.		
1617:15 CAM-3	so---		
1617:15 CAM-1	I think the stab trim thing is broke---		
1617:17 CAM-3	---I didn't wanna call you guys... but---		
1617:18 CAM-1	no no that's good.		
1617:20 CAM-3	---that girl, they're like you better go up there---		

Appendix B 228 **Aircraft Accident Report**

INTRA-COCKPIT COMMUNICATION		AIR-GROUND COMMUNICATION	
TIME and SOURCE	CONTENT	TIME and SOURCE	CONTENT
1617:21 CAM-1	I need you everybody strapped in now, dear.		
1617:22 CAM-3	---and tell them.		
1617:23 CAM-3	ok.		
1617:24 CAM-1	cause I'm gonna I'm going to release the back pressure and see if I can get it... back.		
1617:30 CAM	[sound similar to cockpit door operating]		
1617:33 CAM-2	three zero one eight.		
1617:37 CAM-1	I'll get it here.		
1617:40 CAM-2	I don't think you want any more speedbrakes do you?		
1617:42 CAM-1	uhh no. actually.		
1617:46 CAM-2	he wants us to maintain seventeen.		
1617:51 CAM-1	ok I need help with this here.		

Appendix B   229   Aircraft Accident Report

INTRA-COCKPIT COMMUNICATION

TIME and SOURCE	CONTENT
1617:52 CAM-1	slats ext... lets----
1617:54 CAM-2	ok slats----
1617:54 CAM-1	gimme slats extend.
1617:55 CAM-2	got it.
1617:56.6 CAM	[sound similar to slat/flap handle movement]
1617:58 CAM-1	I'm test flyin now----
1617:59 CAM-2	how does it feel?
1618:00 CAM-1	it's wantin to pitch over more on you.
1618:02 CAM-2	really?
1618:03 CAM-1	yea.
1618:04 CAM-2	try flaps?.... fifteen, eleven?
1618:05 CAM-1	ahh lets go to eleven.

AIR-GROUND COMMUNICATION

CONTENT

Appendix B 230 **Aircraft Accident Report**

INTRA-COCKPIT COMMUNICATION		AIR-GROUND COMMUNICATION	
TIME and SOURCE	**CONTENT**	**TIME and SOURCE**	**CONTENT**
1618:07.3 CAM	[sound similar to slat/flap handle movement]		
1618:09 CAM-2	ok... get some power on.		
1618:10 CAM-1	I'm at two hundred and fifty knots, so I'm lookin.....		
1618:17 CAM-2	real hard?		
1618:17 CAM-1	no actually its pretty stable right here... see but we got to get down to a hundred an eighty.		
1618:26 CAM-1	OK... bring bring the flaps and slats back up for me.		
1618:32 CAM-2	slats too?		
1618:33 CAM-1	yea.		
1618:36.8 CAM	[sound similar to slat/flap handle movement]		
1618:37 CAM-2	that gives us... twelve thousand pounds of fuel, don't over boost them.		
1618:47 CAM-1	what I'm what I wanna do...		

Appendix B 231 **Aircraft Accident Report**

INTRA-COCKPIT COMMUNICATION		AIR-GROUND COMMUNICATION	
TIME and SOURCE	CONTENT	TIME and SOURCE	CONTENT
1618:48 CAM	[sound similar to slat/flap handle movement]		
1618:49 CAM-1	is get the nose up... and then let the nose fall through and see if we can stab it when it's unloaded.		
1618:54 CAWS	[sound of chime] Altitude (repeats for approximately 34 seconds)		
1618:56 CAM-2	you mean use this again? I don't think we should... if it can fly, its like----		
1619:01 CAM-1	it's on the stop now, its on the stop.		
1619:04 CAM-2	well not according to that its not.		
1619:07 CAM-2	the trim might be, and then it might be uh, if something's popped back there----		
1619:11 CAM-1	yea.		
1619:11 CAM-2	----it might be * mechanical damage too.		
1619:14 CAM-2	I think if it's controllable, we oughta just try to land it----		

Appendix B   232   Aircraft Accident Report

INTRA-COCKPIT COMMUNICATION		AIR-GROUND COMMUNICATION	
TIME and SOURCE	**CONTENT**	**TIME and SOURCE**	**CONTENT**
1619:16 CAM-1	you think so? ok lets head for L.A.		
1619:21.1 CAM	[sound of faint thump]		
1619:24 CAM-2	you feel that?		
1619:25 CAM-1	yea.		
1619:29 CAM-1	ok gimme sl--- see, this is a bitch.		
1619:31 CAM-2	is it?		
1619:31 CAM-1	yea.		
1619:32.8 CAM	[sound of two clicks similar to slat/flap handle movement]		
1619:36 CAM-?	*		
1619:36.6 CAM	[sound of extremely loud noise] [increase in background noise begins and continues to end of recording] [sound similar to loose articles moving around in cockpit]		
1619:37 CAM-?	*		

INTRA-COCKPIT COMMUNICATION

TIME and SOURCE	CONTENT
1619:37.6 PA	[sound similar to CVR startup tone]
1619:43 CAM-2	mayday.
1619:49 CAM-1	push and roll, push and roll.
1619:54 CAM-1	ok, we are inverted... and now we gotta get it....
1619:59 CAM	[sound of chime]
1620:03 CAM-1	kick *
1620:04 CAM-1	push push push... push the blue side up.
1620:14 CAM-1	push.
1620:14 CAM-2	I'm pushing.
1620:16 CAM-1	ok now lets kick rudder... left rudder left rudder.
1620:18 CAM-2	I can't reach it.
1620:20 CAM-1	ok right rudder... right rudder.

AIR-GROUND COMMUNICATION

TIME and SOURCE	CONTENT

INTRA-COCKPIT COMMUNICATION		AIR-GROUND COMMUNICATION	
TIME and SOURCE	CONTENT	TIME and SOURCE	CONTENT
1620:25 CAM-1	are we flyin?... we're flyin... we're flyin... tell 'em what we're doin.		
1620:33 CAM-2	oh yea let me get *		
1620:35 CAM-1	*		
1620:38 CAM-1	gotta get it over again... at least upside down we're flyin.		
1620:40.6 PA	[sound similar to CVR startup tone]		
1620:42 CAM-?	*		
1620:44 CAM-?	*		
1620:49 CAM	[sounds similar to compressor stalls begin and continue to end of recording]		
1620:49 CAM	[sound similar to engine spool down]		
1620:54 CAM-1	speedbrakes.		
1620:55.1 CAM-2	got it.		

Appendix B — Aircraft Accident Report

INTRA-COCKPIT COMMUNICATION

TIME and SOURCE	CONTENT
1620:56.2 CAM-1	ah here we go.
1620:57.1	[end of recording]

End of transcript

AIR-GROUND COMMUNICATION

TIME and SOURCE	CONTENT

this page intentionally left blank

www.ingramcontent.com/pod-product-compliance
Lightning Source LLC
Chambersburg PA
CBHW080239180526
45167CB00006B/2338